Construction Planning, Equipment and Methods

Construction Planning, Equipment and Methods

Naseer Khan

RANDOM PUBLICATIONS
NEW DELHI (INDIA)

Construction Planning, Equipment and Methods

ISBN 978-93-5111-429-1

Published in 2014 in India by

RANDOM PUBLICATIONS

4376-A/4B, Gali Murari Lal, Ansari Road
New Delhi-110 002
Phone : +91-11-43580356, +91-11-23289044
e-mail: randomexports@gmail.com, sales@randompublications.com,
info@randompublications.com

Reprinted 2023

Type Setting by: Friends Media, Delhi-110089
Digitally Printed at : Replika Press Pvt. Ltd.

Preface

Construction planning is a fundamental and challenging activity in the management and execution of construction projects. It involves the choice of technology, the definition of work tasks, the estimation of the required resources and durations for individual tasks, and the identification of any interactions among the different work tasks. A good construction plan is the basis for developing the budget and the schedule for work. Developing the construction plan is a critical task in the management of construction, even if the plan is not written or otherwise formally recorded. In addition to these technical aspects of construction planning, it may also be necessary to make organizational decisions about the relationships between project participants and even which organizations to include in a project. For example, the extent to which sub-contractors will be used on a project is often determined during construction planning.

Like a detective, a planner begins with a result (i.e. a facility design) and must synthesize the steps required to yield this result. Essential aspects of construction planning include the generation of required activities, analysis of the implications of these activities, and choice among the various alternative means of performing activities. In contrast to a detective discovering a single train of events, however, construction planners also face the normative problem of choosing the best among numerous alternative plans. Moreover, a detective is faced with an observable result, whereas a planner must imagine the final facility as described in the plans and specifications.

In developing a construction plan, it is common to adopt a primary emphasis on either cost control or on schedule control. Some projects are primarily divided into expense categories with associated costs. In these cases, construction planning is cost or expense oriented. Within the categories of expenditure, a distinction is made between costs incurred directly in the performance of an activity and indirectly for the accomplishment of the project. For example, borrowing expenses for project financing and overhead items are commonly treated as indirect costs. For other projects, scheduling of work activities over time is critical and is emphasized in the planning process. In this case, the planner insures that the proper precedences among activities are maintained and that efficient scheduling of the available resources prevails. Traditional scheduling procedures emphasize the maintenance of task precedences or efficient use of resources over time. Finally, most complex projects require consideration of both cost and scheduling over time, so that planning, monitoring and record keeping must consider both dimensions. In these cases, the integration of schedule and budget information is a major concern.

This book 'Construction Planning, Equipment, And Methods' is a reference book that explains concepts to its readers in a simple, logical and concise format. The book also explains machine capabilities and how it is applied to construction challenges.

I thank all members of my team who have helped in the preparation of the book. My special thanks go to "Random Publications" who have published the book.

—Naseer Khan

Contents

1

History of Structural Engineering

The history of structural engineering dates back to at least 2700 BC when the step pyramid for Pharaoh Djoser was built by Imhotep, the first engineer in history known by name. Pyramids were the most common major structures built by ancient civilizations because it is a structural form which is inherently stable and can be almost infinitely scaled (as opposed to most other structural forms, which cannot be linearly increased in size in proportion to increased loads).

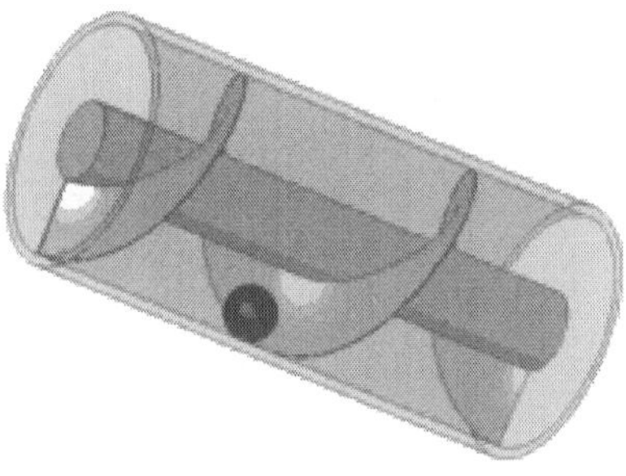

Figure: *The Archimedes screw was operated by hand and could raise water efficiently.*

Another notable engineering feat from antiquity still in use today is the qanat water management system. Qanat technology developed in the time of the Medes, the predecessors of the Persian Empire (modern-day Iran which has the oldest and longest Qanat (older than 3000 years and longer than 71 km) that also spread to other cultures having had contact with the Persian.

Throughout ancient and medieval history most architectural design and construction was carried out by artisans, such as stone masons and carpenters, rising to the role of master builder. No theory of structures existed and understanding of how structures stood up was extremely limited, and based almost entirely on empirical evidence of 'what had worked before'. Knowledge was retained by guilds and seldom supplanted by advances. Structures were repetitive, and increases in scale were incremental.

No record exists of the first calculations of the strength of structural members or the behaviour of structural material, but the profession of structural engineer only really took shape with the industrial revolution and the re-invention of concrete. The physical sciences underlying structural engineering began to be understood in the Renaissance and have been developing ever since.

Early Structural Engineering Developments

The recorded history of structural engineering starts with the ancient Egyptians. In the 27th century BC, Imhotep was the first structural engineer known by name and constructed the first known step pyramid in Egypt. In the 26th century BC, the Great Pyramid of Giza was constructed in Egypt. It remained the largest man-made structure for millennia and was considered an unsurpassed feat in architecture until the 19th century AD.

The understanding of the physical laws that underpin structural engineering in the Western world dates back to the 3rd century BC, when Archimedes published his work *On the Equilibrium of Planes* in two volumes, in which he sets out the *Law of the Lever*, stating:

> *"Equal weights at equal distances are in equilibrium, and equal weights at unequal distances are not in equilibrium but incline towards the weight which is at the greater distance."*

Archimedes used the principles derived to calculate the areas and centres of gravity of various geometric figures including triangles, paraboloids, and hemispheres. Archimedes work on this and his work on calculus and geometry, together with Euclidean geometry, underpin much of the mathematics and understanding of structures in modern structural engineering. The ancient Romans made great bounds in structural engineering, pioneering large structures in masonry and concrete, many of which are still standing today. They include aqueducts, thermae, columns, lighthouses, defensive walls and harbours. Their methods are recorded by Vitruvius in his De Architecture written in 25 BC, a manual of civil and structural engineering with extensive sections on materials and machines used in construction. One reason for their success is their accurate surveying techniques based on the dioptre, groma and chorobates.

Centuries later, in the 15th and 16th centuries and despite lacking beam theory and calculus, Leonardo da Vinci produced many engineering designs based on scientific observations and rigour, including a design for a bridge to span the Bosporus. Though dismissed at the time, the design has since been judged to be both feasible and structurally valid.

The foundations of modern structural engineering were laid in the 17th century by Galileo Galilei, Robert Hooke and Isaac Newton with the publication of three great scientific works. In 1638 Galileo published *Dialogues Relating to Two New Sciences*, outlining the sciences of the strength of materials and the motion of objects (essentially defining gravity as a force giving rise to a constant acceleration). It was the first establishment of a scientific approach to structural engineering, including the first attempts to develop a theory for beams. This is also regarded as the beginning of structural analysis, the mathematical representation and design of building structures. This was followed in 1676 by Robert Hooke's first statement of Hooke's Law, providing a scientific understanding of elasticity of materials and their behaviour under load.

Eleven years later, in 1687, Sir Isaac Newton published *Philosophiae Naturales Principle Mathematics*, setting out his Laws of Motion, providing for the first time an understanding of the fundamental laws governing structures. Also in the 17th century, Sir Isaac Newton and Gottfried Leibniz both independently developed the Fundamental theorem of calculus, providing one of the most important mathematical tools in engineering.

Further advances in the mathematics needed to allow structural engineers to apply the understanding of structures gained through the work of Galileo, Hooke and Newton during the 17th century came in the 18th century when Leonhard Euler pioneered much of the mathematics and many of the methods which allow structural engineers to model and analyse structures. Specifically, he developed the Euler-Bernoulli beam equation with Daniel Bernoulli (1700–1782) circa 1750 - the fundamental theory underlying most structural engineering design. Daniel Bernoulli, with Johann (Jean) Bernoulli (1667–1748), is also credited with formulating the theory of virtual work, providing a tool using equilibrium of forces and compatibility of geometry to solve structural problems. In 1717 Jean Bernoulli wrote to Pierre

Varignon explaining the principle of virtual work, while in 1726 Daniel Bernoulli wrote of the "composition of forces".

In 1757 Leonhard Euler went on to derive the Euler buckling formula, greatly advancing the ability of engineers to design compression elements.

Modern Developments in Structural Engineering

Throughout the late 19th and early 20th centuries, materials science and structural analysis underwent development at a tremendous pace.

Though elasticity was understood in theory well before the 19th century, it was not until 1821 that Claude-Louis Navier formulated the general theory of elasticity in a mathematically usable form. In his *leçons* of 1826 he explored a great range of different structural theory, and was the first to highlight that the role of a structural engineer is not to understand the final, failed state of a structure, but to prevent that failure in the first place. In 1826 he also established the elastic modulus as a property of materials independent of the second moment of area, allowing engineers for the first time to both understand structural behaviour and structural materials.

Towards the end of the 19th century, in 1873, Carlo Alberto Castigliano presented his dissertation "Intorno ai sistemi elastici", which contains his theorem for computing displacement as partial derivative of the strain energy.

In 1824, Portland cement was patented by the engineer Joseph Aspdin as *"a superior cement resembling Portland Stone"*, British Patent no. 5022. Although different forms of cement already existed (Pozzolanic cement was used by the Romans as early as 100 B.C. and even earlier by the ancient Greek and Chinese civilizations) and were in common usage in Europe from the 1750's, the discovery made by Aspdin used commonly available, cheap materials, making concrete construction an economical possibility.

Developments in concrete continued with the construction in 1848 of a rowing boat built of ferrocement - the forerunner of modern reinforced concrete - by Joseph-Louis Lambot. He patented his system of mesh reinforcement and concrete in 1855, one year after W.B. Wilkinson also patented a similar system. This was followed in 1867 when a reinforced concrete planting tub was patented by Joseph Monier in Paris, using steel mesh reinforcement similar to that used by Lambot and Wilkinson. Monier took the idea forward, filing several patents for tubs, slabs and beams, leading eventually to the Monier system of reinforced structures, the first use of steel reinforcement bars located in areas of tension in the structure.

Steel construction was first made possible in the 1850's when Henry Bessemer developed the Bessemer process to produce steel. He gained patents for the process in 1855 and 1856 and successfully completed the conversion of cast iron into cast steel in 1858. Eventually mild steel would replace both wrought iron and cast iron as the preferred metal for construction.

During the late 19th century, great advancements were made in the use of cast iron, gradually replacing wrought iron as a material of choice. Ditherington Flax Mill in Shrewsbury, designed by Charles Bage, was the first building in the world with an interior iron frame. It was built in 1797. In 1792 William Strutt had attempted to build a fireproof mill at Belper in Derby (Belper West Mill), using cast iron columns and timber beams within the depths of brick arches that formed the floors. The exposed beam soffits were protected against fire by plaster. This mill at Belper was the world's first attempt to construct fireproof buildings, and is the first example of fire engineering. This was later improved upon with the construction of Belper North Mill, a collaboration between Strutt and Bage, which by using a full cast iron frame represented the world's first "fire proofed" building.

The Forth Bridge was built by Benjamin Baker, Sir John Fowler and William Arrol in 1889, using steel, after the original design for the bridge by Thomas Bouch was rejected following the collapse of his Tay Rail Bridge. The Forth Bridge was one of the first major uses of steel, and a landmark in bridge design. Also in 1889, the wrought-iron Eiffel Tower was built by Gustave Eiffel and Maurice Koechlin, demonstrating the potential of construction using iron, despite the fact that steel construction was already being used elsewhere.

During the late 19th century, Russian structural engineer Vladimir Shukhov developed analysis methods for tensile structures, thin-shell structures, lattice shell structures and new structural geometries such as hyperboloid structures. Pipeline transport was pioneered by Vladimir Shukhov and the Branobel company in the late 19th century.

Again taking reinforced concrete design forwards, from 1892 onwards François Hennebique's firm used his patented reinforced concrete system to build thousands of structures throughout Europe. Thaddeus Hyatt in the US and Wayss & Freitag in Germany also patented systems. The firm *AG für Monierbauten* constructed 200 reinforced concrete bridges in Germany between 1890 and 1897 The great pioneering uses of reinforced concrete however came during the first third of the 20th century, with Robert Maillart and others furthering of the understanding of its behaviour. Maillart noticed that many concrete bridge structures were significantly cracked, and as a result left the cracked areas out of his next bridge design - correctly believing that if the concrete was cracked, it was not contributing to the strength. This resulted in the revolutionary Salginatobel Bridge design. Wilhelm Ritter formulated the truss theory for the shear design of reinforced concrete beams in 1899, and Emil Mörsch improved this in 1902. He went on to demonstrate that treating concrete in compression as a linear-elastic material was a conservative approximation of its behaviour. Concrete design and analysis has been progressing ever since, with the development of analysis methods such as yield line theory, based on plastic analysis of concrete (as opposed to linear-elastic), and many different variations on the model for stress distributions in concrete in compression. Prestressed concrete, pioneered by Eugène Freyssinet with a patent in 1928, gave a novel approach in overcoming the weakness of concrete structures in tension. Freyssinet constructed an experimental prestressed arch in 1908 and later used the technology in a limited form in the Plougastel Bridge in France in 1930. He went on to build six prestressed concrete bridges across the Marne River, firmly establishing the technology.

Structural engineering theory was again advanced in 1930 when Professor Hardy Cross developed his Moment distribution method, allowing the real stresses of many complex structures to be approximated quickly and accurately.

In the mid 20th century John Fleetwood Baker went on to develop the plasticity theory of structures, providing a powerful tool for the safe design of steel structures.

High-rise construction, though possible from the late 19th century onwards, was greatly advanced during the second half of the 20th century. Fazlur Khan designed structural systems that remain fundamental to many modern high rise constructions and which he employed in his structural designs for the John Hancock Centre in 1969 and Sears Tower in 1973. Khan's central innovation in skyscraper design and construction was the idea of the "tube" and "bundled tube" structural systems for tall buildings. He defined the framed tube structure as "a three dimensional space structure composed of three, four, or possibly more frames, braced frames, or shear walls, joined at or near their edges to form a vertical tubelike structural system capable of resisting lateral forces in any direction by cantilevering from the foundation." Closely spaced interconnected exterior columns form the tube. Horizontal loads, for example wind, are supported by the structure as a whole. About half the exterior surface is available for windows. Framed tubes allow fewer interior columns, and so create more usable floor space. Where larger openings like garage doors are required, the tube frame must be interrupted, with transfer girders used to maintain structural integrity. The first building to apply the tube-frame construction was in the DeWitt-Chestnut Apartment Building which Khan designed in Chicago. This laid the foundations for the tube structures used in most later skyscraper constructions, including the construction of the World Trade Centre.

Another innovation that Fazlur Khan developed was the concept of X-bracing, which reduced the lateral load on the building by transferring the load into the exterior columns. This allowed for a reduced need for interior columns thus creating more floor space, and can be seen in the John Hancock Centre. The first sky lobby was also designed by Khan for the John Hancock Centre in 1969. Later buildings with sky lobbies include the World Trade Centre, Petronas Twin Towers and Taipei 101.

In 1987 Jörg Schlaich and Kurt Schafer published the culmination of almost ten years of work on the strut and tie method for concrete analysis - a tool to design structures with discontinuities such as corners and joints, providing another powerful tool for the analysis of complex concrete geometries.

In the late 20th and early 21st centuries the development of powerful computers has allowed finite element analysis to become a significant tool for structural analysis and design. The development of finite element programs has led to the ability to accurately predict the stresses in complex structures, and allowed great advances in structural engineering design and architecture. In the 1960's and 70's computational analysis was used in a significant way for the first time on the design of the Sydney Opera House roof. Many modern structures could not be understood and designed without the use of computational analysis. Developments in the understanding of materials and structural behaviour in the latter part of the 20th century have been significant, with detailed understanding being developed of topics such as fracture mechanics, earthquake engineering, composite materials, temperature effects on materials, dynamics and vibration control, fatigue, creep and others. The depth and breadth of knowledge now available in structural engineering, and the increasing range of different structures and the increasing complexity of those structures has led to increasing specialisation of structural engineers.

A civil engineer is a person who practices civil engineering; the application of planning, designing, constructing, maintaining, and operating infrastructures while protecting the public and environmental health, as well as improving existing infrastructures that have been neglected. Originally, a civil engineer worked on public works projects and was contrasted with the military engineer, who worked on armaments and defences. Over time, various branches of engineering have become recognised as distinct from civil engineering, including chemical engineering, mechanical engineering, and electrical engineering, while much of military engineering has been absorbed by civil engineering. In some places, a civil engineer may perform land surveying; in others, surveying is limited to construction surveying, unless an additional qualification is obtained. On some U.S. Military bases, the personnel responsible for building and grounds maintenance, such as grass mowing, are called civil engineers and are not required to meet any minimum educational requirements.

Specialisation

Civil engineers usually practice in a particular speciality, such as construction engineering, geotechnical engineering, structural engineering, land development, transportation engineering, hydraulic engineering, and environmental engineering. Some civil engineers, particularly those working for government agencies, may practice across multiple specialisation, particularly when involved in critical infrastructure development or maintenance.

Education and Licensing

In most countries, a civil engineer will have graduated from a post-secondary school with a degree in civil engineering, which requires a strong background in mathematics and the physical sciences; this degree is typically a bachelor's degree, though many civil engineers study further to obtain masters, engineer, doctoral and post doctoral degrees. In many countries, civil engineers are subject to licensure. In jurisdictions with mandatory licensing, people who do not obtain a license may not call themselves "civil engineers."

Europe

Belgium: In Belgium, *Civil Engineer* (abbreviated ir.) (French: *Ingénieur Civil*, Dutch: *Burgerlijk Ingenieur*) is a legally protected title applicable to graduates of the five-year engineering course of one of the six universities and the Royal Military Academy. Their speciality can be all fields of engineering: civil, structural, electrical, mechanical, chemical, physics and even computer science. This use of the title may cause confusion to the English speaker as the Belgian "civil" engineer can have a speciality other than civil engineering. In fact, Belgians use the adjective "civil" as an opposition to military engineers.

The formation of the civil engineer has a strong mathematical and scientific base and is more theoretical in approach than the practical oriented industrial engineering. Educated in a five-year program at a polytechnic. Traditionally, students were required to pass an entrance exam on mathematics to start civil engineering studies. This exam was abolished in 2004 for the Flemish Community, but is still organised in the French Community.

Scandinavia

In Scandinavian countries, civil engineer (civilingenjör (Swedish), sivilingeniør (Norwegian), civilingeniør (Danish)) is a first professional degree, approximately equivalent to Master of Science in Engineering, and a protected title granted to students by selected institutes of technology. As in English the word has its origin in the distinction between civilian and military engineers, as in before the start of the 19th century only military engineers existed and the prefix "civil" was a way to separate those who had studied engineering in a regular University from their military counterparts. Today the degree spans over all fields within engineering, like civil engineering, computer science, electronics engineering, etc.

There is generally a slight difference between a Master of Science in Engineering degree and the Scandinavian civil engineer degree, the latter's programme having closer ties with the industry's demands. A civil engineer is the most well-known of the two; still, the area of expertise remains obfuscated for most of the public. A noteworthy difference is the mandatory courses in mathematics and physics, regardless of the equivalent master degree, e.g. computer science. Although a 'college engineer' (högskoleingenjör, diplomingenjör/ mellaningenjör (Swedish), høgskoleingeniør (Norwegian), diplomingeniør (Danish)) is roughly equivalent to a Bachelor of Science in Scandinavia, to become a 'civil engineer' one often has had to do up to one extra year of overlapping studies compared to attaining a B.Sc./M.Sc. combination. This is because the higher educational system is not fully adopted to the international standard graduation system, since it is treated as a professional degree. Today (2009) this is starting to change due to the Bologna process. A Scandinavian "civilingenjör" will in international contexts commonly call herself "Master of Science in Engineering" and will occasionally wear an engineering class ring. At the Norwegian Institute of Technology (now the Norwegian University of Science and Technology), the tradition with a NTH Ring goes back to 1914, before the Canadian iron ring. In Norway the title "Sivilingeniør" will no longer be issued after 2007, and have been replaced with "Master i technology". In the English translation of the diploma, the title will be "Master of Science", since "Master of Technology" is not an established title in the English-speaking world. The extra overlapping year of studies have also been abolished with this change to make Norwegian degrees more equal to their international counterparts.

Spain

In Spain, a civil engineering degree can be obtained after four years of study in the various branches of mathematics, physics, mechanics, etc. The earned degree is called *Grado en Ingeniería Civil.* Further studies at a Graduate school include master's and doctoral degrees. Before the current situation, that is, before the implementation of Bologna Process in 2010, a Civil Engineering degree in Spain could be obtained after three or five years of study. In the first case, the earned degree was called *Ingeniero Técnico de Obras Públicas* (*ITOP*), literally translated as "Public Works Engineer"; at the second case, the academic earned degree was called *Ingeniero de Caminos, Canales y Puertos* (often shortened to *Ingeniero de Caminos* or *ICCP*), that literally means "Roads, Canals and Harbours Engineer", though civil engineers in Spain practice in the same fields as civil engineers do elsewhere.

The first Spanish Civil Engineering School was the *Escuela Especial de Ingenieros de Caminos y Canales* (now called *Escuela Técnica Superior de Ingenieros de Caminos, Canales y Puertos*), established in 1802 in Madrid, followed by the *Escuela Especial de Ayudantes de Obras Públicas* (now called *Escuela Universitaria de Ingeniería Técnica de Obras Públicas de la Universidad Politécnica de Madrid*), founded in 1854 in Madrid. Both schools now belong to the Technical University of Madrid.

In Spain, a Civil Engineer has the technical and legal ability to design projects of any branch, so any Spanish Civil Engineer can oversee projects about structures, buildings (except residential structures which are reserved for architects), foundations, hydraulics, the environment, transportation, urbanism, etc. In Spain, Mechanical and Electrical engineering tasks are included under the Industrial engineering degree.

United Kingdom

A chartered civil engineer (known as certified or professional engineer in other countries) is a member of the Institution of Civil Engineers, and has also passed chartership exams. However a non-chartered civil engineer may be a member of the Institution of Civil Engineers or the Institution of Civil Engineering Surveyors. The description "Civil Engineer" is not restricted to members of any particular professional organisation although "Chartered Civil Engineer" is.

North America

United States: In the United States, civil engineers are typically employed by municipalities, construction firms, consulting engineering firms, architect/engineer firms, state governments, and the federal government. Each State requires engineers who offer their services to the public to be licensed by the State. Licensure is obtained by meeting specified education, examination, and work experience requirements. Specific requirements vary by State. Typically licensed engineers must graduate from an ABET-accredited University or College engineering program, pass the Fundamentals of Engineering exam, obtain several years of engineering experience under the supervision of a licensed engineer, then pass the Principles and Practice of Engineering Exam. After completing these steps and the granting of licensure by a State Board, engineers may use the title "Professional Engineer" or PE in advertising and documents.

Professional Associations

ASCE: The ASCE (American Society of Civil Engineers) represents more than 140,000 members of the civil engineering profession worldwide. Official members of the ASCE must hold a bachelor's degree from an accredited civil engineering program and be a licensed professional engineer or have five years responsible charge of engineering experience. Most civil engineers join this organisation in order to be updated of current news, projects, and methods (such as sustainability) related to civil engineering; as well as contribute their expertise and knowledge to other civil engineers and students obtaining their civil engineering degree.

ICE: The ICE (Institution of Civil Engineers) founded in 1818, represents, as of 2008, more than 80,000 members of the civil engineering profession worldwide. Its commercial arm, Thomas Telford Ltd, provides training, recruitment, publishing and contract services.

Sub-disciplines of Civil Engineering

In general, civil engineering is concerned with the overall interface of human created fixed projects with the greater world. General civil engineers work closely with surveyors and specialised civil engineers to fit and serve fixed projects within their given site, community and terrain by designing grading, drainage, pavement, water supply, sewer service, electric and communications supply, and land divisions.

General engineers spend much of their time visiting project sites, developing community consensus, and preparing construction plans. General civil engineering is also referred to as site engineering, a branch of civil engineering that primarily focuses on converting a tract of land from one usage to another. Civil engineers typically apply the principles of geotechnical engineering, structural engineering, environmental engineering, transportation engineering and construction engineering to residential, commercial, industrial and public works projects of all sizes and levels of construction.

Materials Science and Engineering

One of the major aspects of Civil engineering is materials science. Material engineering deals with ceramics such as concrete, mix asphalt concrete, strong metals such as aluminum and steel, and polymers such as polymethylmethacrylate (PMMA) and carbon fibers.

Materials engineering also consists of protection and prevention like paints and finishes. Alloying is another aspect of materials engineering, combining two types of metals to produce a stronger metal. It incorporates elements of applied physics and chemistry. With significant media attention focused on nanoscience and nanotechnology in recent years, materials science has been propelled to the forefront at many universities. It is also an important part of forensic engineering and failure analysis. Materials science also deals with fundamental properties and characteristics of materials.

Coastal Engineering

Coastal engineering is concerned with managing coastal areas. In some jurisdictions the terms sea defense and coastal protection are used to mean, respectively, defense against flooding and erosion. The term coastal defense is the more traditional term, but coastal management has become more popular as the field has expanded to include techniques that allow erosion to claim land.

Coastal engineering, as it relates to harbours, starts with the development of ancient civilizations together with the origin of maritime traffic, perhaps before 3500 B.C. Docks, breakwaters, and other harbour works were built by hand and often in a grand scale.

Some of the harbour works are still visible in a few of the harbours that exist today, while others have recently been explored by underwater archaeologists. Most of the grander ancient harbour works have disappeared following the fall of the Roman Empire.

Most ancient coastal efforts were directed to port structures, with the exception of a few places where life depended on coastline protection. Venice and its lagoon is one such case. Protection of the shore in Italy, England and the Netherlands can be traced back at least to the 6th century. The ancients understood such phenomena as the Mediterranean currents and wind patterns and the wind-wave cause-effect link.

The Romans introduced many revolutionary innovations in harbour design. They learned to build walls underwater and managed to construct solid breakwaters to protect fully exposed harbours. In some cases wave reflection may have been used to prevent silting. They also used low, water-surface breakwaters to trip the waves before they reached the main breakwater. They became the first dredgers in the Netherlands to maintain the harbour at Velsen. Silting problems here were solved when the previously sealed solid piers were replaced with new open-piled jetties. The Romans also introduced to the world the concept of the holiday at the coast.

Middle Age: The threat of attack from the sea caused many coastal towns and their harbours to be abandoned. Other harbours were lost due to natural causes such as rapid silting, shoreline advance or retreat, etc. The Venetian Lagoon was one of the few populated coastal areas with continuous prosperity and development where written reports document the evolution of coastal protection works. Engineering and scientific skills remained alive in the east, in Byzantium, where the Eastern Roman Empire survived for six hundred years while Western Rome decayed.

Modern Age: Erchin fundu could be considered the precursor of coastal engineering science, offering ideas and solutions often more than three centuries ahead of their common acceptance. Although great strides were made in the general scientific arena, little improvement was done beyond the Roman approach to harbour construction after the Renaissance. In the early 19th century, the advent of the steam engine, the search for new lands and trade routes, the expansion of the British Empire through her colonies, and other influences, all contributed to the revitalisation of sea trade and a renewed interest in port works.

Twentieth Century: Evolution of shore protection and the shift from structures to beach nourishment. Prior to the 1950's, the general practice was to use hard structures to protect against beach erosion or storm damages. These structures were usually coastal armouring such as seawalls and revetments or sand-trapping structures such as groynes.

During the 1920's and 30's, private or local community interests protected many areas of the shore using these techniques in a rather ad hoc manner. In certain resort areas, structures had proliferated to such an extent that the protection actually impeded the recreational use of the beaches. Erosion of the sand continued, but the fixed back-beach line remained, resulting in a loss of beach area.

The obtrusiveness and cost of these structures led in the late 1940's and early 1950's, to move toward a new, more dynamic, method. Projects no longer relied solely on hard coastal defence structures, as techniques were developed which replicated the protective characteristics of natural beach and dune systems.

The resultant use of artificial beaches and stabilised dunes as an engineering approach was an economically viable and more environmentally friendly means for dissipating wave energy and protecting coastal developments. Over the past hundred years the limited knowledge of coastal sediment transport processes at the local authorities level has often resulted in inappropriate measures of coastal erosion mitigation. In many cases, measures may have solved coastal erosion locally but have exacerbated coastal erosion problems at other locations -up to tens of kilometres away- or have generated other environmental problems.

Planning Approaches

There are five generic strategies for coastal defence:

- inaction leading to eventual abandonment
- Managed retreat or realignment, which plans for retreat and adopts engineering solutions that recognise natural processes

of adjustment, and identifies a new line of defence where to construct new defences

- Hold the line, shoreline protection, whereby seawalls are constructed around the coastlines
- Move seawards, this happens by constructing new defences seaward the original ones
- Limited intervention, accommodation, by which adjustments are made to be able to cope with inundation, raising coastal land and buildings vertically.

The decision to choose a strategy is site-specific, depending on pattern of relative sea-level change, geomorphological setting, sediment availability and erosion, as well a series of social, economic and political factors. Alternatively, integrated coastal zone management approaches may be used to prevent development in erosion- or flood-prone areas to begin with. Growth management can be a challenge for coastal local authorities who often struggle to provide the infrastructure required by new residents seeking sea change lifestyles. Sustainable transport investment to reduce the average footprint of coastal visitors is often a good way out of coastal gridlock. Examples include Dongtan and the Gold Coast Oceanway.

Managed Retreat

In the context of coastal erosion, managed retreat (also managed realignment) allows an area that was not previously exposed to flooding by the sea to become flooded by removing coastal protection. This process is usually in low lying estuarine areas and almost always involves flooding of land that has at some point in the past been claimed from the sea. Managed retreat is often a response to sea level rise exacerbated by local subsidence of the land surface due to isostatic rebound in the north. In the UK the main reason for implementation of Managed Realignment is generally to improve coastal stability, essentially replacing artificial 'hard' coastal defences with natural 'soft' coastal landforms (Pethick 2002). This process can be used to protect areas of land further inland rather than that near the coast by relying on natural defences to absorb or dampen the force of waves.

In addition to being used as a means of coastal defence, Managed Realignment has also been used in a number of cases to mitigate for loss of intertidal habitat. Although land claim has been an important factor for salt marsh loss in the UK in the past (Allen 1992) the majority of current salt marsh loss in the UK is believed to be due to erosion (Morris et al. 2004). This erosion may involve coastal squeeze, where protective sea walls prevent the landward migration of salt marsh in response to sea level rise when sediment supply is limited (Hulme 2005; Morris et al. 2004). Salt marshes are protected under the EU Habitats Directive as well as providing habitat for a number of species protected by the Birds Directive. Following this guidance, the UK's biodiversity action plan aims to prevent net losses to the area of salt marsh present in 1992. It is therefore a legal requirement that all losses in marsh area must be compensated by replacement habitat with equivalent biological characteristics (Crooks et al. 2001). This equates to the need to restore approximately 1.4 km^2 of salt marsh habitat per year in the UK.

Advantages

There are no direct costs apart from that of removing any defences already in place and maintenance costs are very low.

Sediment flow is also restored to its natural state, beaches can be naturally replenished due to erosion of the coast, providing protection and the balance of the coastline returns.

Disadvantages

A certain amount of land will inevitably be lost in this process while beaches are being built up resulting in settlements, farmland and other property being destroyed. Because of this, managed retreat is often not a socially acceptable plan and may invoke the need for compensation to landowners. Intertidal sites are often a rich archaeological resource and the loss of heritage is a factor to weighed in managed retreat projects.

There are no agreed protocols on the monitoring of MR sites (Atkinson et al. 2001) and, consequently, very few of the sites are being monitored consistently and effectively (Wolters et al. 2005c). Due to the low levels of monitoring there is little evidence on which to base future managed realignment projects. This has led to the results of Managed Realignment schemes being extremely unpredictable.

Current Progress

At present approximately 6 km^2 of salt marsh have been restored by MR in the UK (Mossman et al. In prep). One of the major reasons cited for the slow pace of current salt marsh restoration in the UK (Morris et al. 2004) is the uncertainty associated with the practice (Foresight).

Hold the Line

Human strategies on the coast have been heavily based on a static engineered response, whereas the coast is in, or strives towards, a dynamic equilibrium (Schembri, 2009). Solid coastal structures are built and persist because they protect expensive properties or infrastructures, but they often relocate the problem downdrift or to another part of the coast. Soft options like beach nourishment, while also being temporary and needing regular replenishment, appear more acceptable, and go some way to restore the natural dynamism of the shoreline. However in many cases there is a legacy of decisions that were made in the past which have given rise to the present threats to coastal infrastructure and which necessitate immediate shore protection. For instance, the seawall and promenade of many coastal cities in Europe represents a highly engineered use of prime seafront flange-eating space, which might be preferably designated as public open space, parkland and amenities if it were available today. Such open space might also allow greater flexibility in terms of future land-use change, for instance through managed retreat, in the face of threats of erosion or inundation as a result of sea-level rise.

Foredunes areas represent a natural reserve which can be called upon in the face of extreme events; building on these areas leaves little option but to undertake costly protective measures when extreme events (whether amplified by gradual global change or not) threaten.

Managed retreat can comprise 'setbacks', rolling easements and other planning tools including building within a particular design life. Maintenance of those structures or soft techniques can arrive at a critical point (economically or environmental) to change adopted strategy.

- Structural or hard engineering techniques, i.e. using permanent concrete and rock constructions to "fix" the coastline and protect the assets locate behind. These techniques—seawalls, groynes, detached breakwaters, and revetments—represent a significant share of protected shoreline in Europe (more than 70%).
- Soft engineering techniques (e.g. sand nourishments), building with natural processes and relying on natural elements such as sands, dunes and vegetation to prevent erosive forces from reaching the backshore. These techniques include beach nourishment and sand dune stabilisation.

Move Seaward

The futility of trying to predict future scenarios where there is a large human influence is apparent. Even future climate is to a certain extent a function of what humans choose to make of it, for example by restricting greenhouse gas emissions to control climate change. In some cases - where new areas are needed for new economic or ecological development - a move seaward strategy can be adopted. Some examples from erosion are Koge Bay (Dk) Western Scheldt estuary (NI), Chatelaillon (F), Ebro delta (E)

There is an obvious downside to this strategy. Coastal erosion is already widespread, and there are many coasts where exceptional high tides or storm surges result in encroachment on the shore, impinging on human activity. If the sea rises, many coasts that are developed with infrastructure along or close to the shoreline will be unable to accommodate erosion, and will experiment a so-called "coastal squeeze". This occurs where the ecological or geomorphological zones that would normally retreat landwards encounter solid structures and are squeezed out. Wetlands, salt marshes, mangroves and adjacent fresh water wetlands are particularly likely to suffer from this squeeze.

An upside to the strategy is that moving seaward (and upward) can create land of high value which can bring the investment required to cope with climate change.

Limited Intervention

Limited intervention is an action taken whereby the management only solves the problem to some extent, usually in areas of low economic significance. Measures taken using limited intervention often encourage the succession of haloseres, including salt marshes and sand dunes. This will normally result in the land behind the halosere being more sufficiently protected, as wave energy will be dissipated by the accumulated sediment and additional vegetation residing in the newly formed habitat.

Although the new halosere is not strictly man-made, as many natural processes will contribute to the succession of the halosere, anthropogenic factors are partially responsible for the formation as an initial factor was needed to help start the process of succession. This must not be confused with 'accommodate' which is about property e.g. effective insurance, early warning systems and not about habitat.

Construction Techniques

The following is a catalogue of relevant techniques that could be employed as coastal management techniques. *The costs given are very rough estimates made during 2005, based on UK Pound sterling.*

Groyne

A groyne (groin in the United States) is a rigid hydraulic structure built from an ocean shore (in coastal engineering) or from a bank (in rivers) that interrupts water flow and limits the movement of sediment. In the ocean, groynes create beaches, or avoid having them washed away by longshore drift.

In a river, groynes prevent erosion and ice-jamming, which in turn aids navigation. Ocean groynes run generally perpendicular to the shore, extending from the upper foreshore or beach into the water. All of a groyne may be under water, in which case it is a submerged groyne. The areas between groups of groynes are groyne fields. Groynes are generally made of wood, concrete, or rock piles, and placed in groups. They are often used in tandem with seawalls.

Groynes, however, may cause a shoreline to be perceived as unnatural and unattractive.

In Coastal Engineering

A groyne's length and elevation, and the spacing between groynes is determined according to local wave energy and beach slope. Groynes that are too long or too high tend to accelerate downdrift erosion because they trap too much sediment. Groynes that are too short, too low, or too permeable are ineffective because they trap too little sediment. Flanking may occur if a groyne does not extend far enough landward.

How Groynes Work

A groyne creates and maintains a wide area of beach or sediment on its updrift side, and reduces erosion on the other. It is a physical barrier to stop sediment transport in the direction of longshore transport (also called Longshore Drift). This causes a buildup, which is often accompanied by accelerated erosion of the downdrift beach, which receives little or no sand from longshore drift (this is known as terminal groyne syndrome, as it occurs after the terminal groyne in a group of groynes). Groynes do not add extra material to a beach, but merely retain some of the existing sediment on the updrift side of the groynes. If a groyne is correctly designed, then the amount of material it can hold will be limited, and excess sediment will be free to move on through the system. However, if a groyne is too large it may trap too much sediment, which can cause severe beach erosion on the downdrift side.

In Rivers: River groynes (spur dykes or wing dykes) (American English: "dikes") are often constructed nearly perpendicular to the riverbanks, beginning at a riverbank with a root and ending at the regulation line with a head. They maintain a channel to prevent ice jamming, and more generally improve navigation and control over lateral erosion, that would form from meanders. Groynes have a major impact on the river morphology: they cause autonomous degradation of the river.

They are also used around bridges to prevent bridge scour.

Types

Groynes can be distinguished by how they are constructed, whether they are submerged, their effect on stream flow or by shape.

By Construction Method: Groynes can be permeable, allowing the water to flow through at reduced velocities, or impermeable, blocking and deflecting the current.

- Permeable groynes are large rocks, bamboo or timber
- impermeable groynes (solid groynes or rock armour groynes) are constructed using rock, gravel, gabions.

By Whether they are Submerged: Groynes can be submerged or not under normal conditions. Usually impermeable groynes are non-submerged, since flow over the top of solid groynes may cause severe erosion along the shanks. Submerged groynes, on the other hand, may be permeable depending on the degree of flow disturbance needed.

By their Effect on Stream Flow: Groynes can be attracting, deflecting or repelling.

- Attracting groynes point downstream, serving to attract the stream flow toward themselves and not repel the flow toward the opposite bank. They tend to maintain deep current close to the bank.
- Deflecting groynes change the direction of flow without repelling it. They are generally short and used for limited, local protection.
- Repelling groynes point upstream; they force the flow away from themselves. A single groyne may have one section, for example, attracting, and another section deflecting.

By Shape: Groynes can be built with different plan view shapes. Examples are straight groynes, T head, L head, hockey stick, inverted hockey stick groynes, straight groynes with pier head, wing, and tail groynes.

Seawall

A seawall (also written as sea wall) is a form of coastal defence constructed where the sea, and associated coastal processes, impact directly upon the landforms of the coast. The purpose of a seawall is to protect areas of human habitation, conservation and leisure activities from the action of tides and waves. As a seawall is a static feature it will conflict with the dynamic nature of the coast and impede the exchange of sediment between land and sea. The coast is generally a high-energy, dynamic environment with spatial variations occurring over a wide range of temporal scales. The shoreline is part of the coastal interface which is exposed to a wide range of erosional processes arising from fluvial, aoelian and terrestrial sources, meaning that a combination of denudational processes will work against a seawall. Given the natural forces that seawalls are constantly subjected to, maintenance (and eventually replacement) is an ongoing requirement if they are to provide an effective long term solution.

The many types of seawall in use today reflects both the varying physical forces they are designed to withstand, and location specific aspects, such as: local climate, coastal position, wave regime, and value of landform. Seawalls are classified as a hard engineering shore based structure used to provide protection and to lessen coastal erosion. However, a range of environmental problems and issues may arise from the construction of a seawall, including disrupting sediment movement and transport patterns, which are discussed in more detail below. Combined with a high construction cost, this has led to an increasing use of other soft engineering coastal management options such as beach replenishment. Seawalls may be constructed from a variety of materials, most commonly: reinforced concrete, boulders, steel, or gabions. Additional seawall construction materials may include: vinyl, wood, aluminium, fibreglass composite, and with large biodegrable sandbags made of jute and coir. In the UK, *sea wall* also refers to an earthen bank used to create a polder, or a dike.

Design Principles and Types

A seawall works by reflecting incident wave energy back into the sea, therefore reducing the energy and erosion which the coastline would otherwise be subjected to. In addition to their unsightly visual appearance, two specific weaknesses of seawalls exist. Firstly, wave reflection induced by the wall may result in scour and subsequent lowering of the sand level of the fronting beach. Secondly, seawalls may accelerate erosion of the adjacent, unprotected coastal properties because they affect the littoral drift process. Fundamentally, a cost-benefit approach is an effective way to determine whether a seawall is appropriate or not and if the negative effects are worth the protection of threatened property. The design and type of a seawall varies depending on unique aspects specific to each location, and the erosional processes and environment which they are placed in. There are three main types of seawalls: vertical; curved or stepped; and mounds. *Vertical seawalls :* are built in particularly exposed situations. These reflect wave energy and, under storm conditions, standing waves (clapotis) will develop. In some cases piles are placed in front of the wall to lessen wave energy slightly. Curved seawalls are designed to enable waves to break to dissipate wave energy and to repel waves back to the sea. The curve can also prevent the wave overtopping the wall and provides additional protection for the toe of the wall.

Mound-type structures : (revetments, riprap) are used in less demanding settings where lower energy erosional processes operate. The least exposed sites involve the lowest-cost bulkheads and revetments of sand bags or geotextiles. These serve to armour the shore and minimise erosion and may be either watertight or porous, which allows water to filter through after the wave energy has been dissipated.

History

Seawall construction was first documented in 1623 in Canvey Island, UK, when great floods of the Thames estuary occurred, prompting the construction of protection for further events in this flood prone area (Council of Europe, 1999). Since then, seawall design has become more complex and intricate in response to an improvement in materials, technology and an understanding of how coastal processes operate. This section will outline some key case studies of seawalls in chronological order and describe how they have performed in response to tsunami or ongoing natural processes and how effective they were in these situations. Analysing the successes and shortcomings of seawalls during severe natural events allows their weaknesses to be exposed, and areas become visible for future improvement and reassessment.

Pondicherry: On December 26, 2004, towering waves of the 2004 Indian Ocean earthquake tsunami crashed against India's southeastern coastline killing thousands. However, the former French colonial enclave of Pondicherry (now Puducherry) escaped unscathed. This was primarily due to French engineers who had constructed (and maintained) a massive stone seawall during the time which the city was a French colony. This 300 year old seawall effectively kept Pondicherry's historic centre dry even though tsunami waves drove water 24 feet above the normal high-tide mark. The barrier was initially completed in 1735 and over the years, the French continued to fortify the wall, piling huge boulders along its 1.25 mile (2 km) coastline to stop erosion from the waves pounding the harbour. At its highest, the barrier running along the water's edge reaches about 27 feet above sea level. The boulders, some weighing up to a ton, are weathered black and brown. The sea wall is inspected every year and whenever gaps appear or the stones sink into the sand, the government adds more boulders to keep it strong (Allsop, 2002). The Union Territory of Pondicherry recorded some 600 deaths from the huge tsunami waves that struck India's coast after the mammoth underwater earthquake (which measured 9.0 on the moment magnitude scale) off Indonesia, but most of those killed were fishermen who lived in villages beyond the artificial barrier which reinforces the effectiveness of seawalls.

Vancouver Seawall

The Vancouver Seawall is a stone seawall constructed around the perimetre of Stanley Park in Vancouver. The seawall was constructed initially as waves created by ships passing through the First Narrows were eroding the area between Prospect Point and Brockton Point. The Vancouver Seawall also exemplifies how seawalls can be utilised and valued for recreational activities and coastal sightseeing. A pedestrian, cycling and roller blading pathway exists on the seawall and has been extended far outside the parametres of Stanley Park. Construction of the seawall began in 1914, and since then this pathway has become one of the most used features of the park by both locals and tourists and now extends 22 km in total (Belyea & Ross, 1992). The construction of the seawall also provided employment for relief workers during the Great Depression and seamen from the HMCS Discovery on Deadman's Island who were facing punishment detail in the 1950's (Steele, 1985). Overall, the Vancouver Seawall is a prime example of how seawalls can simultaneously provide shoreline protection and a source of recreation which enhances human enjoyment of the coastal environment. It also illustrates that although shoreline erosion is a natural process, human activities, interactions with the coast and poorly planned shoreline development projects can accelerate natural erosion rates.

Japan: At least 40 per cent of Japan's 35,000 kilometre coastline is lined with concrete seawalls or other structures designed to protect the country against high waves, typhoons or even tsunamis (New York Times, 2011). When a Tsunami struck in 2011 following a magnitude 9 offshore earthquake, the seawalls in most areas were overwhelmed. In Kamaishi, 4 metre waves surmounted the seawall —the world's largest, erected a few years ago in the city's harbour at a depth of 63 metres, a length of 2 kilometres and a cost of \$1.5 billion — and eventually submerged the city centre (Musubi, 2011).

The risks of dependence on seawalls was most evident in the crisis at the Daiichi and Daini nuclear power plants, both located along the coast close to the earthquake zone, as the tsunami washed over walls that were supposed to protect the plants. Arguably, the additional defence provided by the seawalls presented an extra margin of time for citizens to evacuate and also stopped some of the full force of energy which would have caused the wave to climb higher in the backs of coastal valleys. In contrast, the seawalls also acted in a negative way to trap water and delay its retreat.

The failure of the world's largest seawall, which cost \$1.5 billion to construct, shows that building stronger sea walls to protect larger areas would have been too costly to be effective. In the case of the ongoing crisis at the nuclear power plants, higher and stronger sea walls should have been built if power plants were to be built at that site. Fundamentally, the devastation in coastal areas and a final death toll predicted to exceed 10,000 could push Japan to redesign its seawalls or consider more effective alternative methods of coastal protection for extreme events. Such hardened coastlines can also provide a false sense of security to property owners and local residents as evident in this situation (Msubi, 2011).

Future Research

Sea Level Rise

Sea level rise creates an issue for seawalls worldwide as it raises both the mean normal water level and the height of waves during extreme weather events, which the current seawall heights may be unable to cope with (Allan *et al.* 1999).

The International Panel on Climate Change (IPCC) (1997) suggested that sea level rise over the next 50 – 100 years will accelerate with a projected increase in global mean sea level of +18 cm by 2050 AD. This data is reinforced by Hannah (1990) who calculated similar statistics including a rise of between +16-19.3 cm throughout 1900–1988. This problem could be overcome by further modelling and determining the extension of height and reinforcement of current seawalls which needs to occur for safety to be ensured in both situations.

Extreme Events

Extreme events also pose a problem as it is not easy for people to predict or imagine the strength of hurricane or storm induced waves compared to normal, expected wave patterns. An extreme event can dissipate hundreds of times more energy than everyday waves, and calculating structures which will stand the force of coastal storms is difficult and, often the outcome can become unaffordable. For example, Omaha Beach seawall in New Zealand was designed to prevent erosion from everyday waves only, and when a storm in 1976 carved out 10 m behind the existing seawall the whole structure was destroyed (GeoResources, 2001).

Other Limitations

Some further limitations include: lack of long term trend data of seawall effects due to a relatively short duration of data records; modelling limitations and comparisons of different projects and their effects being invalid or unequal due to different beach types; materials; currents; and environments (Christchurch City Council, 2009).

Revetment

Revetments, or revêtements (the original French word, meaning something to recloth or recover), have a variety of meanings in architecture, engineering and art history. In stream restoration, river engineering or coastal management, they are sloping structures placed on banks or cliffs in such a way as to absorb the energy of incoming water. In military engineering they are structures, again sloped, formed to secure an area from artillery, bombing, or stored explosives. In architecture they are a variety of structures, normally vertical, used to retain a wall, or sometimes just to decorate it. River or coastal revetments are usually built to preserve the existing uses of the shoreline and to protect the slope, as defence against erosion.

Freshwater Revetments

Many revetments are used to line the banks of freshwater rivers, lakes, and man-made reservoirs, especially to prevent damage during periods of floods or heavy seasonal rains. Many materials may be used: wooden piles, loose-piled boulders or concrete shapes, or more solid banks.

Revetments are used as a low-cost solution for coastal erosion defence in areas where crashing waves may otherwise deplete the coastline. Wooden revetments are made of planks laid against wooden frames so that they disrupt the force of the water. Although once popular, the use of wooden revetments has largely been replaced by modern concrete-based defence structures such as tetrapods.

Architecture and Art History

Revetment can be used as a term for a retaining wall, or just for the covering of a wall. In particular the term is used for stone slabs or decorated ceramic plaques used as the outer facing layer of a wall, especially in Ancient Roman architecture. These may or may not have a structural function in the internal construction of the wall, and may be essentially decorative. Marble or terracotta was used, the latter often decorated in moulded reliefs. Revetment is also a term used for a riza or decorated metal cover for most, typically all but the face and perhaps hands, of an Eastern Orthodox icon. The wooden revetments are made of thin wooden planks against other sticks of wood. They are not the best of coastal erosion preventing resources but can still help with easing coastal erosion. The water passes through the wooden planks making the water not crash against the cliff face or any other face.

Rock Armour (Riprap)

Riprap — also known as rip rap, rubble, shot rock or rock armour or "Rip-rap" — is rock or other material used to armour shorelines, streambeds, bridge abutments, pilings and other shoreline structures against scour, water or ice erosion.

It is made from a variety of rock types, commonly granite or limestone, and occasionally concrete rubble from building and paving demolition. It is used to protect coastlines and structures from erosion by the sea, rivers, or streams. It can be used on any waterways or water containment where there is potential for water erosion.

Protection Mechanism

Riprap works by absorbing and deflecting the energy of waves before they reach the defended structure. The size and mass of the riprap material absorbs the impact energy of waves, while the gaps between the rocks trap and slow the flow of water, lessening its ability to erode soil or structures on the coast. The mass of riprap also provides protection against impact damage by ice or debris, which is particularly desirable for bridge supports and pilings. It is frequently used to protect the base of old Edwardian and Victorian sea walls, which being vertical are often undermined. The riprap absorbs the impact of the waves as they shoot up the wall and then fall back down. In the Western United States, *riprap* can also mean a cross between cobblestones and stairs.

Gabion

Gabions are cages, cylinders, or boxes filled with soil or sand that are used in civil engineering, road building, and military applications. For erosion control caged riprap is used. For dams or foundation construction, cylindrical metal structures are used. In a military context, earth or sand-filled gabions are used to protect artillery crews from enemy fire.

Leonardo da Vinci designed a type of gabion called a *Corbeille Leonard* ("Leonard[o] basket") for the foundations of the San Marco Castle in Milan.

Civil Engineering

In civil engineering a gabion wall is a retaining wall made of rectangular containers (baskets) fabricated of thick galvanised wire, which are filled with stone and stacked on one another, usually in tiers that step back with the slope rather than vertically.

The most common civil engineering use of gabions is to stabilize shorelines or slopes against erosion. Other uses include retaining walls, temporary floodwalls, to filter silt from runoff, for small or temporary/permanent dams, river training, channel lining. They may be used to direct the force of a flow of flood water around a vulnerable structure. Gabions are also used as fish barriers on small streams.

Gabion baskets have some advantages over loose riprap because of their modularity and ability to be stacked in various shapes; they are also resistant to being washed away by moving water. Gabions also have advantages over more rigid structures because they can conform to ground movement, dissipate energy from flowing water, and drain freely. Their strength and effectiveness may increase with time in some cases, as silt and vegetation fill the interstitial voids and reinforce the structure. They are sometimes used to keep stones which may fall from a cutting or cliff from endangering traffic on a thoroughfare.

As all methods of construction, gabions have an approximated life expectancy, and in the case of gabions it relies on the lifespan of the wire and not on the contents of the basket as the structure will fail when the wire fails. In one scenario PVC-coated galvanised gabions have been estimated to survive for 60 years. Some gabion manufacturers guarantee a structural consistency of 50 years. Although galvanised steel wire is most widely used, stainless steel wire gabions are also developed and used.

Figure: *Bridge abutment with gabions.*

Gabions have also been used in building, as in the Dominus Winery in Napa Valley, California. The exterior is formed by modular wire mesh gabions containing locally quarried stone; this construction creates an environment of moderate temperatures within the building.

In the medieval era, gabions were round cages with open tops and bottoms, made from wickerwork and filled with earth for use as military fortifications. These early military gabions were used to protect field artillery gunners. The wickerwork cylinders were light and could be carried relatively conveniently in the ammunition train, particularly if they were made in several diametres to fit one inside another. At the site of use in the field, they could be stood on end, staked in position, and filled with soil to form an effective wall around the gun. Today, gabions are often used to protect forward operating bases (FOBs) against explosive, fragmentary, indirect fires such as mortar or artillery fire. Examples of areas within a FOB that make extensive use of gabions would be sleeping quarters, mess halls, or any place where there will be a large concentration of unprotected soldiers.

Cliff Stabilisation

Cliff stabilisation is a coastal management erosion control technique. This is most suitable for softer or less stable cliffs. Generally speaking, the cliffs are stabilised through dewatering (drainage of excess rainwater to reduce water-logging) or anchoring (the use of terracing, planting, or wiring to hold cliffs in place).

Training (Civil)

Training or entrance training refers to coastal structures built to constrain a river discharging across a littoral coast so that it discharges only where desired. Untrained entrances on sandy coasts tend to move widely and violently to discharge into the ocean, often upsetting those enjoying land nearby. With many cities (and buildings) constructed close to rivers, such management has historically been considered a necessary course of action, even though ecologically, nonintervention would be better and more sustainable.

A trained entrance often consists of rock walls that force the water into a deeper more stable channel. Trained entrances can provide better navigation, water quality and flood mitigation services, but can also cause beach erosion due to their interruption of longshore drift. One solution is the installation of a sand bypass system across the trained entrance.

Training is also used on mountainous rivers and streams, and ensures that a fast-flowing river is reduced in violence (and hence erosive capability), usually by the use of weirs and other structures like gabions.

Floodgate

Floodgates are adjustable gates used to control water flow in flood barriers, reservoir, river, stream, or levee systems. They may be designed to set spillway crest heights in dams, to adjust flow rates in sluices and canals, or they may be designed to stop water flow entirely as part of a levee or storm surge system. Since most of these devices operate by controlling the water surface elevation being stored or routed, they are also known as crest gates.

In the case of flood bypass systems, floodgates sometimes are also used to lower the water levels in a main river or canal channels by allowing more water to flow into a flood bypass or detention basin when the main river or canal is approaching a flood stage.

Types

- Bulkhead gates are vertical walls with movable, or removable, sections. Movable sections can be lifted to allow water to pass underneath (as in a sluice gate) and over the top of the structure. Historically, these gates used stacked timbers known as stop logs or wooden panels known as flashboards to set the dam's crest height. Some floodgates known as coupures in large levee systems slide sideways to open for various traffic. Bulkhead gates can also be made of other materials and used as a single bulkhead unit. Mitre gates are used in ship locks and usually close at an 18° angle to approximate an arch.
- Hinged crest gates, are wall sections that rotate from vertical to horizontal, thereby varying the height of the dam. They are generally controlled with hydraulic power, although some are passive and are powered by the water being impounded.

  Variations;
  - flap gate
  - fish-belly flap gates
  - Bascule gates
  - Pelican gates.
- Radial gates are rotary gates consisting of cylindrical sections. They may rotate vertically or horizontally. Tainter gates are a vertical design that rotates up to allow water to pass underneath. Low friction trunnion bearings, along with a face shape that balances hydrostatic forces, allow this design to close under its own weight as a safety feature.
- Drum gates are hollow gate sections that float on water. They are pinned to rotate up or down. Water is allowed into or out of the flotation chamber to adjust the dam's crest height.
- Roller gates are large cylinders that move in an angled slot. They are hoisted with a chain and have a cogged design that interfaces with their slot.
- Clamshell gates have an external clamshell leaf design.
- Fusegates are an innovative spillway control technology, which consists of free standing blocks (the Fusegates) set side by side

on a flattened spillway sill. The Fusegate blocks act as a fixed weir most of the time and operate independently without any remote control or energy source only in case of excessive flood conditions. The System is developed and patented by Hydroplus from Paris, France. It has been installed on more than 50 dams around the world with sizes ranging from 1m to more than 9 m in height. Fusegate are typically used to increase the storage capacity of existing dams or to maximize the discharge potential of undersized spillways.

Valves

Valves used in floodgate applications have a variety of design requirements and are usually located at the base of dams. Often, the most important requirement (besides regulating flow) is energy dissipation. Since water is very heavy, it exits the base of a dam with the enormous force of water pushing from above. Unless this energy is dissipated, the flow can erode nearby rock and soil and damage structures.

Other design requirements include taking into account pressure head operation, the flow rate, whether the valve operates above or below water, and the regulation of precision and cost.

- Fixed cone valves, also known as Howell-Bunger valves, are designed to dissipate the energy from a water flow during reservoir discharge. They are a round pipe section with an adjustable sleeve gate and cone at the discharge end. Flow is varied by moving the sleeve away or towards its cone seat. The design allows high pressure water from the base of a dam to be released without causing erosion to the surrounding environment. Fixed cone valves are able to handle heads up to 300 m.
- Hollow jet valves are a type of needle valve used for floodgate discharge. A cone and seat are inside a pipe. Water flows through an annular gap between the pipe and cone when it is moved downstream, away from the seat. Ribs support the bulb assembly and supply air for water jet stabilisation.
- Ring jet valves are similar to fixed cone valves, but have an integral collar that discharges water in a narrow stream. They are suitable for heads up to 50 m.
- Jet flow gate, similar to a gate valve but with a conical restriction prior to the gate leaf that focuses the water into a jet. They were developed in the 1940's by the United States Bureau of

Reclamation to allow fine control of discharge flow without the cavitation seen in regular gate valves. Jet flow gates are able to handle heads up to 150 m.

Physics

In order to do a simple calculation of the force on a rectangular flood gate one can use the following equation:

$$F = pA$$

where:

F = force measured in the SI units kg·m·s^{-2} which is called the newton (N)

p = pressure $= \rho gh$ measured in N/m^2, which is called the pascal (Pa)

where:

- ρ (rho) is the density of fresh water (1000 kg/m^3);
- g is the acceleration due to gravity on Earth (9.8 m/s^2);
- h is the height of the water column in metres.

A = area = rectangle : length × height measured in m^2

where:

length = the horizontal length of a rectangular floodgate measured in metres

height = the height of a non-submerged flood gate from the bottom of the water column to the water surface measured in metres

If the rectangular flood gate is submerged below the surface the same equation can be used but only the height from the water surface to the middle of the gate must be used to calculate the force on the flood gate.

Beach Nourishment

Beach nourishment— also referred to as beach replenishment— describes a process by which sediment (usually sand) lost through longshore drift or erosion is replaced from sources outside of the eroding beach. A wider beach can reduce storm damage to coastal structures by dissipating energy across the surf zone, protecting upland structures and infrastructure from storm surges, tsunamis and unusually high tides. Beach nourishment is typically part of a larger coastal defence scheme. Nourishment is typically a repetitive process, since nourishment does not remove the physical forces that cause erosion; it simply mitigates their effects. A benign environment increases the interval between nourishment projects, reducing costs. Conversely, high erosion rates may render nourishment financially impractical. In many coastal areas, the economic impacts of a wide beach can be substantial. The 10 miles (16 km), long shoreline fronting Miami Beach, Florida was replenished over the period 1976–1981. The project cost approximately $64,000,000 and revitalised the area's economy. Prior to nourishment, in many places the beach was too narrow to walk along, especially during high tide.

The first nourishment project in the U.S. was at Coney Island, New York in 1922-23 and is now a common shore protection measure utilised by public and private entities.

Nourishment is one of three commonly accepted methods for protecting shorelines. The structural alternative involves constructing a seawall, revetment, groin or breakwater. Alternatively, with "managed retreat" the shoreline is left to erode, while relocating buildings and infrastructure further inland. Nourishment gained popularity because it preserved beach resources and avoided the negative effects of hard structures. Instead, nourishment creates a "soft" (i.e., non-permanent) structure by creating a larger sand reservoir, pushing the shoreline seaward.

History

The first nourishment project in the U.S. was constructed at Coney Island, New York in 1922-1923.

Causes of Erosion

Beaches can erode both naturally and due to impact of humans.

Erosion is a natural response to storm activity. During storms, sand from the visible beach submerges to form storm bars that protect the beach. Submersion is only part of the cycle. During calm weather smaller waves return sand from the storm bar to the visible beach surface in a process called accretion. The term *erosion* conjures visions of environmental damage so the term submersion often replaces it in describing a healthy sandy beach. Continental drift erodes coastlines naturally. Ocean currents can change, disrupting the submersion/ accretion cycle.

Some beaches do not have enough sand available to coastal processes to respond naturally to storms. When there is not enough sand left available on a beach, then there is no recovery of the beach following storms.

Many areas of high erosion are due to human activities. Reasons can include seawalls locking up sand dunes, coastal structures like ports and harbour's that prevent longshore drift, dams and other river management structures. These activities interfere with the natural sediment flows either through dam construction (thereby reducing riverine sediment sources) or construction of littoral barriers such as jetties, or by deepening of inlets; thus preventing longshore transport of sediment across these channels.

Visible and Submerged Sand

The distinction between total sand in a beach and the proportion of the sand above the waterline (submersion fraction) critically impacts beach nourishment. Two beaches with the same amount of visible sand may look much different under water. An eroded beach with substantial submerged sand surrounding it may recover without nourishment. Nourishing a beach that has little submerged sand requires addressing the reason that the submerged sand is missing. Otherwise, the same forces that stripped the submerged sand once are likely to do so again. The amount of submerged sand eroded is typically much greater than the amount of missing sand on shore. Replacing only the visible sand is insufficient without replacing the sand that once supported the accretion part of the process is insufficient. In that circumstance, the beach is unstable and the visible sand quickly erodes. If human activity is a major cause of the erosion, mitigating that activity may be more cost effective over both short and long term periods than nourishment.

Requirements for Effective Nourishment

Sediment texture (grain size and sorting) is critical for success. Sand fill must be compatible with native beach sand. In some cases, beaches have been nourished using a finer sand than the original. Thermolumine scence monitoring reveals that storms can erode such beaches far more quickly than the natural beach. This was observed at the Waikiki nourishment project in Hawaii.

Profile Nourishment

Beach Profile Nourishment describes programs that nourish the full beach profile. In this instance, "profile" means the slope of the uneroded beach from above the water to well out to sea, not just the visible portion. The Gold Coast profile nourishment program placed 75% of its total sand volume below low water level. Some coastal authorities *overnourish* the below water beach (aka "nearshore nourishment") so that over time the natural beach increases in size. These approaches do not permanently protect beaches eroded by human activity. Doing so still requires mitigating that activity.

Replenishment Material

The selection of suitable material for a particular project depends upon the design needs, environmental factors, transport costs considering both short and long-term implications.

The most important material characteristic is the sediment's grain size, which must closely match the native material. Excess silt and clay fraction (mud) versus the natural turbidity in the nourishment area disqualifies some materials. Projects that did not match grain sizes performed relatively poorly. Nourishment sand that is only slightly smaller than native sand can result in significantly narrower equilibrated dry beach widths compared to sand the same size as (or larger than) native sand. Evaluating material fit requires a sand survey that usually includes geophysical profiles and surface and core samples.

Nourishment Projects

The setting of a beach nourishment project is key to design and potential performance. Possible settings include a long straight beach, an inlet that may be either natural or modified, and a pocket beach. Rocky or seawalled shorelines, that otherwise have no sediment, present unique problems.

Cancun, Mexico

Federal and state governments in Mexico have invested about $71 million dollars ($957 million pesos) throughout the state of Quintana Roo in restoring the beaches along Cancun, Playa del Carmen, and Cozumel.

Hurricane Wilma hit the beaches of Cancun and the Riviera Maya in 2005. The initial nourishment project was unsuccessful, leading to a second round that began in September 2009, and was scheduled to complete in early 2010. The project designers and the government committed to invest in beach maintenance to address future erosion. Project designers considered factors such as the time of year and sand characteristics such as density. Restoration in Cancun was expected to deliver 1.3 billion US gallons (4,900,000 m^3) of sand to replenish 450 metres (1,480 ft) of coastline. This time, the beach is promised to last at least 10 years.

Northern Gold Coast, Queensland, Australia

Gold Coast beaches in Queensland, Australia have experienced periods of severe erosion. In 1967 a series of 11 cyclones removed most of the sand from Gold Coast beaches. The Government of Queensland engaged engineers from Delft University in the Netherlands to advise them. The 1971 Delft Report outlined a series of works for Gold Coast Beaches, including beach nourishment and an artificial reef. By 2005 most of the recommendations had been implemented.

The Northern Gold Coast Beach Protection Strategy (NGCBPS) was a Aus $10 million dollar investment. NGCBPS was developed between 1992 and 1999 and the works were completed between 1999 and 2003. The project included dredging 3,500,000 cubic metres (4,600,000 cu yd) of compatible sand from the Gold Coast Broadwater and delivering it through a pipeline to nourish 5 kilometres (3.1 mi) between of beach between Surfers Paradise and Main Beach. The new sand was stabilised by an artificial reef constructed at Narrowneck out of huge geotextile sand bags. The new reef was designed to improve wave conditions for surfing. A key monitoring program for the NGCBPS is the ARGUS coastal camera system operated by the University of New South Wales. The cost/benefit ratio for NGCBPS was conservatively estimated at 75:1 for a AUS $10 million investment into beach replenishment. The benefits were estimated from a model of lost visitor nights in hotels following previous erosion events. NGCBPS so improved beach health that recovery following minor and moderate storms occurred within weeks. Additional unquantified benefits included lifestyle benefits for residents, additional public open space and improved fishing, diving and surfing conditions.

Netherlands

More than one-quarter of the Netherlands is below sea level and about 80% of the coast consists of sand dune or beach. The shoreline is closely monitored by yearly recording of the cross section at points 250 metres (820 ft) apart, to ensure adequate protection. Where long-term erosion is identified, beach nourishment using high-capacity suction dredgers is deployed.

Hawaii

Waikiki: Hawaii planned to replenish Waikiki beach in 2010. Budgeted at $2.5 million, the project covered 1,700 feet (520 m) in an attempt to return the beach to its 1985 width. Prior opponents supported this project, because the sand was to come from nearby shoals, reopening a blocked channel and leaving the overall local sand volume unchanged, while closely matching the "new" sand to existing materials. The project planned to apply up to 24,000 cubic yards (18,000 m^3) of sand from deposits located 1,500 to 3,000 feet (460 to 910 m) offshore at a depth of 10 to 20 feet (3.0 to 6.1 m). The project was larger than the prior recycling effort in 2006-07, which moved 10,000 cubic yards (7,600 m^3).

Maui

Maui, Hawaii illustrated the complexities of even small-scale nourishment projects. A project at Sugar Cove transported upland sand to the beach. The sand allegedly was finer than the original sand and contained excess silt that enveloped coral heads, smothering the coral and killing small animals that lived in and around it. As in other projects, onshore sand availability was limited, forcing consideration of more expensive offshore sources.

A second project, along Stable Road, that attempted to slow rather than halt erosion, was stopped halfway toward its goal of adding 10,000 cubic yards (7,600 m^3) of sand. The beaches had been retreating at a "comparatively fast rate" for half a century. The restoration was complicated by the presence of old seawalls, groins, piles of rocks and other structures. 7This project used sand-filled Geotube groins placed that were originally to remain in place for up to 3 years. The pipeline was anchored by concrete blocks attached by fibre straps. A video showed the blocks bouncing off the coral in the current, killing whatever they touched. In places the straps broke, allowing the pipe to move across the reef, "planing it down". Bad weather exacerbated the damaging movement and led to project shutdown. The smooth, cylindrical Geotubes could be difficult to climb over before they were covered by sand.

Supporters claimed that 2010's seasonal summer erosion was less than in prior years, although the beach was narrower after the restoration ended than in 2008. Authorities were studying whether to require the project to remove the groins immediately. Potential alternatives to Geotubes for moving sand included floating dredges and/or trucking in sand dredged offshore. A final consideration was sea level rise and that Maui was sinking under its own weight. Both islands surround massive mountains (Haleakala, Mauna Loa, and Mauna Kea) and were expanding a giant dimple in the ocean floor, some 30,000 feet (9,100 m) below the mountain summits.

Beach Drainage

Beach drainage or beach face dewatering lowers the water table locally beneath the beach face. This causes accretion of sand above the drainage system. Grant (1946) – the elevation of the beach water table had an important bearing on deposition and erosion across the foreshore.

A high water table coincided with periods of accelerated beach erosion, and conversely, a low water table coincided with pronounced aggradation of the foreshore A lower water table (unsaturated beach face) facilitates deposition by reducing flow velocities during backwash and prolonging laminar flow. In contrast, a high water table results in condition favouring beach erosion. With the beach in a saturated state, Grant proposed that backwash velocity is accelerated by the addition of ground water seepage out of the beach within the effluent zone.

Turner and Leatherman (1997) moving from the origins and development of the dewatering concept to field and laboratory studies available at the time of writing concluded that there was too little evidence for being convinced that the systems had a positive effect. None of the case studies provide full scientific evidence of indisputable positive results regarding beach stabilisation although in some cases an overall positive performance was reported. In many cases no adequate long-term monitoring was undertaken at a frequency high enough to discriminate the response to high energy erosive events.

A useful side effect of the system is that the collected seawater is very pure because of the sand filtration effect. It may be discharged back to sea but can also be used to oxygenate stagnant inland lagoons /marinas or used as feed for heat pumps, desalination plants, land based aquaculture, aquariums or seawater swimming pools.

Beach drainage systems have been installed in many locations around the world to halt and reverse erosion trends in sand beaches. Twenty four beach drainage systems have been installed since 1981 in Denmark, USA, UK, Japan, Spain, Sweden, France, Italy and Malaysia.

Costs

The costs of installation and operation per metre of shoreline protection will vary due to:

- system length (non-linear cost elements)
- pump flow rates (sand permeability, power costs)

- soil conditions (presence of rock or impermeable strata)
- discharge arrangement /filtered seawater utilisation
- drainage design, materials selection & installation methods
- geographical considerations (location logistics)
- regional economic considerations (local capabilities /costs)
- study requirements /consent process.

The costs associated with a beach drainage system are generally considerably lower than hard engineered structures. They also compare very favourably with beach nourishment projects, particularly when long-term project economics are considered (nourishment projects often have a limited life or a program of re-nourishment).

Monitoring Coastal Zones

Coastal zone managers are faced with difficult and complex choices about how best to reduce property damage in the shorelines. One of the problems they face is error and uncertainty in the information available to them on the processes that cause erosion of beaches. Video-based monitoring lets collect data continuously at low cost and produce analyses of shoreline processes over a wide range of averaging intervals.

Event Warning Systems

Event warning systems, such as tsunami warnings and storm surge warnings, can be used to minimize the human impact of catastrophic events that cause coastal erosion. Storm surge warnings can also be used to determine when to close floodgates to reduce the physical impact of such events. Wireless sensor networks can be deployed quickly to set up a coastal erosion monitoring system, and scaled accordingly.

Shoreline Mapping

Defining the shoreline is a difficult task due to the dynamic nature of the coast and the intended application of the shoreline (Graham et al. 2003; Boak & Turner 2005). Given this idea the shoreline must therefore be considered in a temporal sense whereby the scale is dependent on the context of the investigation (Boak & Turner 2005).

The following definition of the coast and shoreline is most commonly employed for the purposes of shoreline mapping. The coast comprises the interface between land and sea, and the shoreline is represented by the margin between the two (Woodroffe, 2002). Due to the dynamic nature of the shoreline coastal investigators adopt the use of shoreline indicators to represent the true shoreline position (Boak & Turner 2005).

Shoreline Indicator

The choice of shoreline indictor is a primary consideration in shoreline mapping. According to Leatherman (2003) it is important that indicators are easily identified in the field and on aerial photography. Shoreline indicators may be physical beach morphological features such as the berm crest, scarp edge, vegetation line, dune toe, dune crest and cliff or the bluff crest and toe. Alternatively, non-morphological features may also be used. These indicators are based on water level including the high water line, mean high water line, wet/dry boundary, and the physical water line (Pajak & Leatherman 2000). Figure 1 provides a sketch of the spatial relationship between many of the commonly used shoreline indicators. The high water line (HWL), defined as the wet/dry line is the most commonly used shoreline indicator because it is visible in the field, and can be interpreted on both colour and grey scale aerial photographs (Leatherman, 2003; Crowell et al. 1991). The HWL represents the landward extent of the most recent high tide and is characterised by a change in sand colour due to repeated, periodic inundation by high tides. The HWL is portrayed on aerial photographs by the most landward change in colour or grey tone (Boak & Turner 2005).

Importance and Application

The location of the shoreline and its changing position over time is of fundamental importance to coastal scientists, engineers and managers (Boak & Turner 2005; Pajack & Leatherman 2002). Present day shoreline monitoring campaigns provide information about historic shoreline location and movement, and about predictions of future change (Appeaning Addo et al. 2008). More specifically the position of the shoreline in the past, at present and where it is predicted to be in the future is useful for in the design of coastal protection, to calibrate and verify numerical modells to assess sea level rise, map hazard zones and formulate policies to regulate coastal development. Accurate and consistent delineation of the shoreline is integral to all of these tasks. The location of the shoreline also provides information regarding shoreline reorientation adjacent to structures, beach width, volume and rates of historical change (Boak & Turner 2005; Pajack & Leatherman 2002).

Data Sources

A variety of data sources are available for examining shoreline position however, the availability of historical data is limited at many coastal sites and so the choice of data source is largely limited to what is available for the site at a given time (Boak & Turner 2005). Shoreline mapping techniques applied to data sources have moved towards automation in association with technological advances and the need to reduce uncertainty. Although these changes have resulted in improvement in coastal data processing and storage capabilities, the frequent change in technology has prevented the emergence of one standard method of shoreline mapping. This has occurred because each data source and associated method have their own unique capabilities and shortcomings (Moore 2000). A number of the data sources used for shoreline mapping and their associated advantages and disadvantages.

Historical Maps

In the event that a study requires the shoreline position to be mapped before the development of aerial photographs, or if the location has poor photograph coverage it is necessary to employ historical maps in order to detail shoreline position (Moore 2000). The main advantage and reason for using historical maps is that they are able to provide a historic record that is not available from other data sources. Many potential errors however are associated with historical coastal maps and charts. Such errors may be associated with scale, datum changes, distortions from uneven shrinkage, stretching, creases, tears and folds, different surveying standards, different publication standards, and projection errors (Boak & Turner 2005). The severity of these errors depends on the accuracy standards met by each map and the physical changes that have occurred since the publication of the map (Anders & Byrnes 1991). The oldest reliable source of shoreline data in the United States dates back to the early to mid 19th century and is the U.S Coast and Geodetic Survey/National Ocean Service T-sheets (Morton 1991). In the United Kingdom, many maps and charts were deemed to be inaccurate until around 1750. The founding of the Ordnance Survey in 1791 has since improved the accuracy of the mapping.

Aerial Photographs

Aerial photographs have been used since the 1920's to provide topographical information about an area. They are therefore a good database for compilation of historical shoreline change maps. Aerial photographs are the most commonly used data source in shoreline mapping because many coastal areas have extensive aerial photo coverage therefore providing a valuable record of shoreline position (Moore 2000). In general, aerial photographs provide good spatial coverage of the coast however temporal coverage is very much site specific depending on the flight path of the aeroplane. A second disadvantage associated with aerial photography is that the interpretation of the shoreline position is subjective given the dynamic nature of the coastal environment. This combined with various distortions inherent in aerial photographs can lead to significant error levels (Moore 2000).

Object Space Displacements

Conditions outside of the camera can cause objects in an image to be displaced from their true ground position. Such conditions may include ground relief, camera tilt and atmospheric refraction.

Relief displacement is prominent when photographing a variety of elevations. This situation causes objects above ground level to be displaced outward from the centre of the photograph and objects below ground level to be displaced toward the centre of the image (Figure 2). The severity of the displacement is affected negatively with decreases in flight altitude and as radial distance from the centre of the photograph increases. This distortion can be minimised by photographing numerous swaths and creating a mosaic of the images. This technique will create a focus for the centre of each photograph where distortion is minimised. It is important to note that this error is not common in shoreline mapping is the relief is fairly constant. It is however important to consider when mapping cliffs (Moore 2000).

Ideally aerial photographs are taken so the optical axis of the camera is perfectly perpendicular to the ground surface thereby creating a vertical photograph. Unfortunately this is not often the case and virtually all aerial photographs experience tilt whereby up to 3° is not uncommon (Camfield et al. 1996). In this situation the scale of the image will be larger on the upward side of the tilt axis and smaller on the downward side. Moore, (2000) notes that many coastal researchers have not realised the severity of this error and therefore do not consider it in their methods.

Radial Lens Distortion

Lens distortion varies as a function of radial distance from the isocentre of the photograph meaning that the centre of the image is relatively distortion free, but as the angle of view increases the distortion becomes more prominent. This is a significant source of error in earlier aerial photography but as technology has increased and camera lens have become more refined it has become less of an issue with later photographs. Such a distortion is impossible to correct for without knowing the make and model of the lens used to capture the image. However if overlapping images have been acquired one can digitize the centre portions of the aerial photographs (Crowell et al. 1991).

Delineation of the Shoreline

The dynamic nature of the coast has meant that accurate mapping of an instantaneous shoreline position has been associated with significant uncertainty. This uncertainty arises because at any given time the position of the shoreline is influenced by the short-term effect of the tide and a wide variety of long term effects such as relative sea-level rise and along shore littoral sediment movement. Not only does this affect the accuracy of computed historic shoreline position but also any predicted future positions (Appeaning Addo et al. 2008). As mentioned earlier the HWL is most commonly used as a shoreline indicator. This can usually be seen as a significant tonal change on aerial photographs. There are however many errors associated with using the wet/dry line as a proxy for the HWL and shoreline. The errors of largest concern are the short term migration of the wet/dry line, interpretation of the wet/dry line on a photograph and measurement of the interpreted line position (Leatherman 2003; Moore 2000). Systematic errors such as the migration of the wet/dry line may arise from tidal and seasonal changes.

Storm-induced erosion is another factor which may cause the wet/dry line to migrate landward. Field investigations have shown that these changes can be minimised by using only summertime data (Moore 2000; Leatherman 2003). Furthermore, the error bar can be significantly reduced by using the longest record of reliable data to calculate erosion rates (Leatherman 2003). Finally it is important to note that errors may arise due to the difficulty of measuring a single line on a photograph. For example where the pen line is 0.13 mm thick this translates to an error of ±2.6 m on a 1:20000 scale photograph.

Beach Profiling Surveys

Beach profiling surveys are typically repeated at regular intervals along the coast in order to measure short-term (daily to annual) variations in shoreline position and beach volume. (Smith & Zarillo

1990). Beach profiling is a very accurate source of information however measurements are generally subject to the limitations of conventional surveying techniques. Shoreline data derived from beach profiling is often spatially and temporally limited due to the high cost associated with such a labour intensive activity. Shorelines are generally derived by interpolating between a series of discrete beach profiles. It is important to note however that the distance between the profiles is usually quite large and so the accuracy of the interpolating becomes compromised. In contrast to aerial photographs, survey data is limited to smaller lengths of shoreline generally less than ten kilometres (Boak & Turner 2005). Beach profiling data is commonly available in from regional councils in New Zealand such as those compiled by the Hawkes Bay Regional Council.

Remote Sensing

Technological advancement over the last decade has led to the development of a range of airborne, satellite and land based remote sensing techniques (Smith & Zarillo 1990). Some of the remotely sensed data sources are listed below:

- Multispectral and hyperspectral imaging
- Microwave sensors
- Global positioning system (GPS)
- Airborne light detection and ranging technology (LIDAR).

Remote sensing techniques are attractive as they are cost effective, reduce manual error and remove the subjective approach of conventional field techniques (Maiti et al. 2009). Remote sensing is a relatively new concept and so extensive historical observations are unavailable. Given this idea, it is important that coastal morphology observations are quantified by coupling remotely sensed data with other sources of information detailing historic shoreline position from archived sources (Appeaning Addo et al. 2008).

Video Analysis

Video analysis provides quantitative, cost-effective, continuous and long-term monitoring beaches (Turner et al. 2004). The advancement of coastal video systems over the past 15 years has resulted in the extraction of large amounts of geophysical data from images. Such data includes that about coastal morphology, surface currents and wave parametres. The main advantage of video analysis lies in the ability to reliably quantify these parametres with high resolution and coverage in both space and time. This in particular highlights their potential importance as an effective coastal monitoring system and an aid to coastal zone management (Van Koningsveld et al. 2007). Interesting case studies have been carried out using video analysis. Turner et al. (2004) used a video-based ARGUS coastal imaging system to monitor and quantify the regional-scale coastal response to sand nourishment and construction of the world-first Gold Coast artificial (surfing) reef in Australia. In addition, Smit et al. (2007) demonstrated the added value of high resolution video observations for making short-term predictions of near shore hydrodynamic and morphological processes, at temporal scales of metres to kilometres and days to seasons.

2

Construction Engineering Basics

Construction engineering is a professional discipline that deals with the designing, planning, construction, and management of infrastructures such as highways, bridges, airports, railroads, buildings, dams, and utilities. Construction Engineers are unique such that they are a cross between civil engineers and construction managers. Construction engineers learn the designing aspect much like civil engineers and construction site management functions much like construction managers.

The primary difference between a construction engineer and a construction manager is that the construction engineer has the ability to sit for the Professional Engineer license (PE) whereas a construction manager cannot. At the educational level, construction managers are not as focused on design work as they are on construction procedures, methods, and people management. Their primary concern is to deliver a project on time, with in budget, and of the desired quality.

The difference between a construction engineer and civil engineer is only at the educational level as both disciplines are able to sit for the PE exam giving them the same title of engineer. Civil engineering students concentrate more on the design work, gearing them toward a career as a design professional. This essentially requires them to take a multitude of design courses. Construction engineering students take design courses as well as construction management courses. This allows them to understand both the design functions as well as the building requirements needed to design and build today's infrastructures.

Work Activities

Depending on which career path the construction engineer has chosen to follow, an entry-level design engineer normally provide support to project managers and assist with creating conceptual designs, scopes, and cost estimates for the planning and construction of approved projects. It should be noted that a career in design work does require a professional engineer license (PE). Individuals who pursue this career path are strongly advised to sit for the Engineer In Training exam (EIT) while in college as it takes five years post graduate to obtain the PE license. Entry-level construction manager positions are typical called project engineers or assistant project engineers. They are responsible for preparing purchasing requisitions, processing change orders, preparing monthly budgeting reports, and handling meeting minutes. The construction management position does not necessarily require a PE license; however possessing one does make the individual more marketable, as the PE license allows the individual to sign off on temporary structure designs.

Abilities

Construction engineers are problem solvers, they help create infrastructure that best meets the unique demands of its environment. They must be able to understand infrastructure life cycles and have the perspective to solve technical challenges with clarity and imagination. Therefore individuals should have a strong understanding of math and science, but many other skills are required, including critical and analytical thinking, time management, people management and good communication skills.

Educational Requirements

Individuals looking to obtain a construction engineering degree must first ensure that the program is accredited by EAC or Technology Accreditation Commission (TAC) of the Accreditation Board for Engineering and Technology (ABET). ABET accreditation is assurance that a college or university program meets the quality standards established by the profession for which it prepares its students. In the US there are currently twenty-five programs that exist in the entire country so careful college consideration is advised.

A typical construction engineering curriculum is a mixture of engineering mechanics, engineering design, construction management and general science and mathematics. This usually leads to a Bachelor of Science degree. The B.S. degree along with some design or construction experience is sufficient for most entry level positions. Graduate schools may be an option for those who want to go further in depth of the construction and engineering subjects taught at the undergraduate level. In most cases construction engineering graduates look to either civil engineering, engineering management, or business administration as a possible graduate degree. For authority to approve any final designs of public projects (and most any project), a construction engineer must have a professional engineers (P.E.) license.

Job Prospects

Job prospects for construction engineers generally have a strong cyclical variation. For example, starting in 2008 - continuing until at least 2011 - job prospects have been poor due to the collapse of housing bubbles in many parts of the world. This sharply reduced demand for construction, forced construction professionals towards infrastructure construction and therefore increased the competition faced by established and new construction engineers. This increased competition, and a core reduction in quantity demand is in parallel with a possible shift in the demand for construction engineers due to the automation of many engineering tasks, overall resulting in reduced prospects for construction engineers. In early 2010 the United States construction industry had a 27% unemployment rate, this is nearly three times higher than the 9.7% national average unemployment rate. The construction unemployment rate (including tradesmen) is comparable to the United States 1933 unemployment rate - the lowest point of the Great Depression - of 25%.

Remuneration

The average salary for a civil engineer in the UK depends on the sector, and more specifically the level of experience of the individual. A 2010 survey of the remuneration and benefits of those occupying jobs in construction and the built environment industry showed that the average salary of a civil engineer in the UK is £29,582. The average salary varies depending on experience, for example the average annual salary for a civil engineer with between 3 and 6 years experience is £23,813. For those with between 14 and 20 years experience the average is £38,214.

Highway Engineering

Highway engineering is an engineering discipline branching from civil engineering which involves the design, construction and maintenance of Highway Systems. Highway Engineering become prominent towards the latter half of the 20th Century after World War 2. Standards of highway engineering are continuously being improved. Highway Engineers must take into account future traffic flows, design of highway intersections/interchanges, geometric alignment and design, highway pavement materials and design, structural design of pavement thickness, and pavement maintenance.

History of Highway Engineering

The beginning of road construction could be dated back to when the Romans were around. With the advancement of technology from carriages pulled by two horses to vehicles with power equivalent to 100 horses, road development had to follow suit. The construction of modern highways did not begin until the late 19th to early 20th century. The first research dedicated to highway engineering was initiated in the United Kingdom with the introduction of the Transport Research Laboratory (TRL), in 1930. In the USA, highway engineering became an important discipline with the passing of the Federal Aid Highway Act of 1944, which aimed to connect 90% of cities with a population of 50,000 or more. With constant stress from vehicles which grew larger as time passed, improvements to pavement were needed. With technology out of date, in 1958 the construction of the first motorway in Great Britain (the Preston bypass) played a major role in the development of new pavement technology.

Design policies standards used in the United States are typically based on publications of the American Association of State Highway and Transportation Officials as well as research promulgated by the Transportation Research Board, the Institute of Transportation Engineers, the Federal Highway Administration, and the Department of Transportation.

Bridge Engineering

Bridge is a structure which provides a passage to people, vehicles, railways or pipelines to cross various obstacles to travel. Engineers build bridges over obstacles such as lakes, rivers, canyons, and dangerous roads and railway tracks. Without bridges, people would need boats to cross waterways and would have to travel around canyons and ravines. Bridges range in length from a few metres to several kilometres. They are among the largest structures built by man. The demands on design and on materials are very high. A bridge must be strong enough to support its own weight as well as the weight of the people and vehicles that use it. The structure also must resist various natural occurrences, including earthquakes, strong winds, and changes in temperature. Most bridges have a concrete, steel, or wood framework and an asphalt or concrete roadway on which people and vehicles travel.

Basic Components of a Bridge

The majority of bridges are held up by at least two supports set in the ground. The distance between two adjacent supports is called a span of a bridge. The supports at each end of the bridge are called abutments, and the supports that stand between the abutments are called piers. The total length of the bridge is the distance between the abutments. Most short bridges are supported only by abutments and are known as single-span bridges. Bridges that have one or more piers in addition to the abutments are called multi-span bridges. Most long bridges are multi-span bridges. In special cases, like in a pontoon bridge no piers or abutments are provided. It is supported by pontoons (flat-bottomed boats) or other portable floats. There are three main components of a typical bridge in general, namely:

- Superstructure
- Substructure
- Foundation.

The above three components have the following basic parts:

Superstructure :

a) The main constituent of any superstructure is the decking, which comprises of a slab, girder, trusses, etc.

b) The handrails, crash barriers, kerbs, etc.

The three basic types of bridge spans are shown below. Any of these spans may be constructed using beams, girders or trusses. Arch bridges are either simple or continuous (hinged). A cantilever bridge may also include a suspended span.

Substructure :

a) Abutments at the extreme ends of the bridge.

b) Piers at intermediate supports in case of multiple span bridges.

c) Bearings and pedestals for the decking.

Foundation :

a) These include foundations for both abutments and piers. Foundations may be of the type open, well, pile, etc.

Apart from the above, river training works and the approaches to a bridge also form a part of a bridge works.

Selection of Bridge Site

It may not be possible always to have a wide choice of sites for a bridge. This is particularly so in case of bridges in urban areas and flyovers. For rivers bridges in rural areas, usually a wider choice may be available.

The characteristics of an ideal site for a bridge across a river are:

- A straight reach of the river;
- Steady river flow without serious whirls and cross currents;
- A narrow channel with firm banks;
- Suitable high banks above high flood level on each side;
- Rock or other hard inerodible strata close to the river bed level;
- Economical approaches which should not be very high or long or liable to flank attacks of the river during floods; the approaches should be free from obstacles such as hills, frequent drainage crossing, sacred places, graveyards, or built up areas or troublesome land acquisition;
- Proximity to a direct alignment of the road to be connected;
- Absence of sharp curves in the approaches;
- Absence of expensive river training works; and
- Avoidance of excessive underwater construction.

For selecting a suitable site for a major bridge, the investigating engineer should make a reconnaissance survey for about one km on the upstream side and one km on the downstream side of the proposed bridge site and should journey along the road for about one km on either side of the road from the bridge site in order to form a general impression of the landscape and to decide on the type of structure best suited to the site. Care should be taken to investigate a number of probable alternative sites and then decide on the site which is likely to serve the needs of the bridge at the least cost. A brief description of the reasons for selection of a particular site should be furnished in the investigation report along with salient details of alternative sites investigated and rejected.

When the river to be crossed is a meandering river, the bridge should be located at a nodal point, i.e. the location where the river regime is constant serving as a fulcrum about which the river channels swing laterally. To the extent possible, it is desirable to align the bridge at right angles to the river, i.e. to provide a square crossing. Sometimes, a skew crossing which is included to the centre line of the river at an angle different from a right angle has to be provided in order to avoid costly land acquisition or sharp curves on the approaches. A skew bridge usually poses more difficulties in design, construction and maintenance. The location of the bridge in relation to the alignment of the approaches may be decided as below : (a) For bridges of total length less than 60 m, the alignment of the approach will govern; (b) For bridges of total length between 60 m and 300 m, both the proper alignment of the approaches and the requirements of a good bridge site should be considered together in ascertaining the appropriate site; and © For major bridge over 300 m length, the requirements of a good bridge site will govern the alignment.

Bridge Classification

The types of bridges vary in total length, the length of their spans, and the weight they can support. Before deciding which kind to build at a particular place, engineers determine the length of the structure and of each span. They also consider the maximum load the bridge will carry and the materials available to construct it. Considering various aspects, bridges may be classified in the following ways :

1. *Function or purpose to be performed :* Based on this bridges can be classified as an aqueduct over a river or a canal, a viaduct which may be road or railway over a valley, a pedestrian or a foot bridge, a highway bridge, a railway bridge, a road cum rail bridge, etc.
2. *Construction material used :* Under this heading bridges can be grouped as timber bridges, masonry bridges, steel bridges, reinforced cement concrete bridges, prestressed bridges, composite bridges, etc.
3. *Alignment of bridge :* Based on this bridges can be classified as a straight or a skew bridge.
4. *Location of bridge floor relative to the superstructure :* In accordance with the relative position of bridge floor as compared to the superstructure, bridges can be classified as deck, through, semi-through and suspended bridge.
5. *According to the type of superstructure :* Depending upon the type of superstructure, bridges can be grouped as slab, girder, truss, suspension, cable stayed, arch, drawbridges, etc.
6. *Nature of superstructure action :* Under its category bridges can be classified as portal frame bridges, truss bridges, balanced cantilever bridges and suspension bridges.
7. *Location of Highest Flood Level :* Particularly for a highway bridge, under this category bridges may be classified as a high level or non - submersible bridge and a submersible bridge

depending upon the formation level of the bridge relative to the highest flood level.

8. *Fixed or Movable / Method of clearance for navigation:* In case of navigable channels which don't have permanent and sufficient clear waterway, fixed or movable bridges are used namely high - level fixed, movable bascule, movable swing or transporter bridge.
9. *Type of loadings:* As per the Indian Road Congress (IRC), road bridges and culverts may be classified into Class A, Class B and Class 70 R (tracked or wheeled) according to the loads they are designed to carry.
10. *Span length:* Under this heading bridges are grouped as culverts if span length is less than 8 m, minor bridges for span lengths between 8 m and 30 m, major bridges for span length more than 30 m and long span bridges for span length above 120 m.
11. *Method of connections :* Depending upon the method of connections of various superstructure parts particular in case of steel construction, steel bridges can be classified into pinned connected, welded or riveted bridges.
12. *Degree of Redundancy :* Depending upon the degree of redundancy bridges can be classified as determinate bridges and indeterminate bridges.
13. *Inter span relation :* Under this category bridges can be grouped as simply supported, continuous or cantilever type bridges.
14. *Type of service to be put to :* In accordance with the type of service anticipated and duration of use, bridges can be classified as permanent, temporary, military bridges like pontoon and bailey, etc.
15. *Number of lanes:* In accordance with the number of lanes for which a bridge structure has been constructed, bridges can be classified as single lane, two lanes or three lane bridges, etc.

Choice of Superstructure Type

The basic function of a bridge superstructure is to permit uninterrupted smooth passage of traffic over it and to transmit the loads and to transmit the load and forces to the substructure safely through the bearings. Although it is difficult to stipulate the aesthetic requirements, it should, however, be ensured that the type of superstructure adopted is simple, pleasing to the eye, and blends with the environment. No hard and fast rules can be laid regarding a the economy in cost. The designer should, however, be able to evolve the most economical type of superstructure based on his judgement and experience given the particular conditions prevailing at the particular site at the particular time. The following factors are to be considered while selecting the type of a bridge superstructure.

i. The nature of river or streams
ii. Nature of foundation / founding strata available
iii. The amount and type of traffic
iv. Whether used for navigation purposes
v. Climatic conditions
vi. Hydraulic data
vii. Type of available construction material
viii. Labour available
ix. The available facilities for erections
x. Maintenance provisional
xi. The availability of funds
xii. Time available for construction
xiii. Strategic consideration
xiv. Economic consideration
xv. Aesthetic consideration.

Characteristics and Design Aspects of Various Types of Superstructures

Basic Principle

The superstructure of any bridge must be designed so as to satisfy the geometric and load carrying requirements to be met. The geometric requirements depend on the number and widths of traffic lanes and footpaths for which the deck is to cater to. It also depends on the overall alignment and various horizontal and vertical clearances required above and below the roadway. Once the geometric considerations are decided, the superstructure has to be designed to meet various structural design requirements. These include considerations of strength, stiffness and stability. This first requires estimation of internal forces and moments and displacements which the externally applied forces will cause in the selected scheme and form of the structure and then deciding the final section sizes, and reinforcements (and prestressing, if the structure is prestressed). The former process constitutes 'analysis' and the latter process constitutes 'design'. The analysis is generally done based on elastic behaviour of the structure. The design is then done either on elastic (i.e. working) strength basis or on load-factored (i.e. ultimate) strength basis, also ensuring the serviceability criteria (like limiting the flexural crack-widths, deflections and vibrations). Some structural analysis work is done using empirical formulae. If the structure is very complicated, model analysis is resorted to.

The selection of superstructure type suitable for particular requirements will depend upon the comparative advantages and disadvantages of each type.

There are seven main kinds of bridges: (1) slab / voided slab bridges, (2) girder bridges, (3) arch bridges, (4) cantilever bridges, (5) suspension bridges, (6) cable-stayed bridges, (7) truss bridges and (8) other bridges

Material Used for Constructing Superstructure

The following materials are generally adopted for constructing permanent bridge decks :

a. Reinforced concrete
b. Prestressed concrete
c. Steel
d. Masonry
e. Composite construction using steel and reinforced concrete, or reinforced and prestressed concrete.

Reinforced Concrete Superstructure

The common types of deck sections in this category are :

a. Solid slab
b. Slab and girder (T-beam)
c. Hollow box girder.

The deck arrangements can be simple or continuous spans, or frame, balanced cantilever, free cantilever cast in situ precast or segmental.

Composite and Steel Superstructure

The various types of deck in this category are :

a. Longitudinal and transverse beams, with concrete slab

b. Longitudinal plate girders, transverse beams and concrete slab
c. Longitudinal box girder with concrete slab
d. Steel truss.

The deck arrangement can be simple or continuous spans, or arch or frame. The slab can be non-composite or can be made composite with the longitudinal beams by shear connectors.

Special Superstructures :

a. Orthotropic deck
b. Cable stayed deck
c. Suspension type deck.

Age masonry superstructure are generally of arch type.

The composite type of decks generally involve combination of two dissimilar structural elements. Precast or prefabricated beams are connected to RC slab by special structural elements called shear connectors, acting together as a unit. The beams may be either :

a. Prefabricated structural steel beams/girders or
b. Precast reinforced concrete, or precast prestressed concrete beams.

Comparison of Reinforced Concrete and Prestressed Concrete Superstructures

- In recent years a majority of bridges have been constructed in prestressed concrete. However, reinforced concrete bridges in certain circumstances still, are more economical.
- Prestressed concrete bridges offer a high degree of freedom from cracks, low maintenance, long life; and they are particularly suitable for dynamic loads and vibrations, and offer great resistance to fatigue arising from repetitions of live loading.
- In RC construction, live load normally approximately doubles the stress in steel, while in prestressed concrete this increase is very little. Position of neutral axis in RC decks about which moments of areas from tension and compression side balance remains fixed, but of that in prestressed concrete shifts with variation in applied load.
- The greatest advantage of prestressed concrete for bridges is reduction in weight slender appearance, possibility of precasting

in factory conditions either in segments or even in whole spans, possibility of larger spans and freedom from tension and cracking. For long spans, prestressed concrete decks can be nearly comparable in weight with steel decks. It is not possible to quote a minimum prestressed concrete span which can be most economical, it depends on local conditions.

For short bridges the advantage of using smaller amount of concrete because of prestressing may be unimportant since there is always a certain basic sot which cannot be changed by the use of prestressing. In such cases the usual reinforced concrete construction may be used with equal advantages.

Bridge Substructure and Foundation

The ' substructure ' and ' foundation ' are generally known to be as the ' better half ' of a bridge structure. This arises from the fact that in a number of cases the substructure and foundation prove to be more expensive than the superstructure especially in case of bridges over major rivers where foundations are very deep and the function of the superstructure carrying the design loads safely depends on the support given by the foundation.

Even though the foundations are not seen by people in general or by those using the bridges but great expertise goes into the design and construction of these in varied soil types, deep waters, marine clays, turbulent flowing rivers, etc. The foundations sometimes reach depths as much as 100 m below the bed. Foundations are very important part of the bridge and the strata on which they are founded has to be unyielding and capable of sustaining the loads. The bridge piers and foundations are required to sustain various loads due to traffic, wind, temperature, differential settlement, seismic forces, etc. The protection of the substructure and foundation is also of utmost importance to ensure the safety of the bridge against the scour of the bed due to water current.

The bridge superstructure cannot be conceived in isolation but is required to be visualised as an integral part of superstructure, structurally as well as architecturally. They are required not only to be engineered to withstand the multitude of loads but also are required to be carefully architecture to blend or harmonize with the surroundings from aesthetic point of view. Thus an integral design approach is required to achieve an architecturally pleasing structure without compromising on the structural integrity and economy.

Abutments

There are two basic types of abutments; open type or spill through type and closed type or non-spill through type.

Open type abutments are more economical than closed type since they allow the back fill to spill out in between their columns thereby reducing the active earth pressure. But due to this very reason, the backfill may settle down, should it be washed through under heavy rains. In such a case of deep curtain wall should be provided extending downward from the abutment cap, for at least 1.5 m, below the spill line.

Piers or Bents

The term 'Pier' pertains to an intermediate substructure unit between the abutment. A bent is a pier consisting of one or more columns, in the substructure, unit with or without cap.

Pile Bents

Type are economical but are suitable for stream crossing where floating debris is minimal. They may be competitive in placed waters and in dry beds, where soil is poor and footings laid at shallower depths might settle. The preferable minimum width of cap is 75 cm. Pile bents are unsightly for bridges 6 m or more in height and unsuitable if scour is significant.

Solid Piers: They are located in streams where floating debris and rolling bounders accompany the last current. They are preferably for long spans, and may be supported on spread footings, pile foundations or caissons, as appropriate. A pier wall should have a low thickness/height ratio, but should not be less than 30 cm thick. This is the minimum thickness that should be used with two planes of reinforcing steel.

Multicolumn Bents and Multi Columns

They generally support dry land structures. They normally replace pile bents when spread footings are recommended. They may be supported on either spread footings or pile foundations. Columns may be circular rectangular or variable in section to provide architectural effect. The thickness/height ratio of normal height uniform section columns should be in a range of 1/12 to 1/15 for appearance as well as structural reasons. Unnecessarily thick columns can invite high temperature and shrinkage stresses and very slender ones will require buckling analysis.

Single Column Bents and Single Columns

They can sometimes be used to avoid skewed piers. They are adapted to viaducts over city streets where the location of the column may be restricted.

Foundations

Footings are appropriate where scour is low and soil bearing capacity is enough not to create settlement problems, or if rock is available at a shallow depth. In any case, if subsoil water is encountered footings are feasible only if dewatering is possible. Otherwise piles are preferable, which enables transferring the loads directly to lower and firmer substrata, without requiring dewatering. These are costlier than footings, in general. However, in case of deep scour, rolling bounders, floating debris and strong current, the (generally costlier) caissons become necessary. Caissons are relatively massive and generally suit spans not less than 25 m preferably longer.

Choice of Type of Foundation

The choice of foundation for the bridges depends upon the conditions of soil. Cost of substructure increases with the depth of foundation. Similarly, the cost of superstructure per unit length also increases with the span length. Hence the decision on span lengths to depend upon the ratio of the cost of substructure including the foundation versus the cost of superstructure i.e. the girder and the deck and it can be found that it is most economical when the ratio is one. There may be cases where though the cost of foundations may be more than the cost of superstructure practical difficulties in sinking the foundation and limitations may require a reconsideration for the reduction of the total number of foundations. Navigational considerations also will have an influence upon the span. The choice of span has to be suitably made taking all these into consideration.

Foundations can be broadly classified as :

(i) open foundations;

(ii) pile foundations; and

(iii) well foundations.

Open foundations are suitable for small bridges and bridges which re sited in a location where rock or firm founding soil is available at a shallow depth and, at the same time, the scour in the river is not considerable. The next choice, from considerations of economy, are the pile foundation which are again more suitable for being used in cohesive soils and beds subject to minimum scour. Well foundations have been more popularly used in India in view of the expertise available particularly for major and important bridges.

General Design Procedure for Design of Substructure of A Bridge

I. First of all compute the dead load reaction due to the superstructure and superimposed dead load along with the maximum and minimum live load reactions coming on the pier or abutment under consideration. Apart from this also compute the corresponding moments caused due to these loads about the longitudinal as well as the transverse directions or axis of the bridge. For this the critical case is the maximum possible eccentric transfer of these dead and live load reactions along with the braking and temperature forces, etc. to the foundation base.

II. Compute the total vertical reaction at the base of the foundation or at the soft of the pile cap in case of pile foundation under maximum, minimum and no live load conditions, taking into account the upward force due to the effect of buoyancy equal to 50% or 100% weight of water in volume equal to that of the submerged mass. This value is 100% in case of saturated soil and fissured, weathered or weak rock and 50% in case of good rock.

III. Compute the total moment about the bridge's longitudinal axis and the horizontal force in the transverse direction, at various levels and more importantly at the base of the foundation due to possible eccentricity of live and dead loads, force due to water current and afflux, if any.

IV. Compute the total moment about the bridge's transverse axis and the horizontal force in the longitudinal direction, at various levels and more importantly at the base of the foundation due to possible eccentricity of live and dead loads, force due to braking and temperature, water current and afflux, if any. The force due to braking can be calculated depending upon the following cases : (i) In case of simply supported spans on rocker / roller bearings, braking force due to live load on one complete span may be assumed to go to its rocker bearing only such that the corresponding foundation under it will resist force either equal to braking - temperature or half of braking + temperature, the latter case being critical for usually an abutment or a pier supporting unequal spans. (ii) In case of simply supported spans with identical neoprene or elastomeric bearings under each of an individual span, the foundation will resist the sum of half of the value of braking force due to live loads on each of the two spans supported by it and the horizontal temperature force equal to the difference of the product of the shear rating of the neoprene bearings and the deck movements above them for left and right hand side spans respectively. (iii) In case of continuous bridge structures the horizontal force being resisted by the bearings depends upon the arrangement and type of bearings chosen and is case specific.

V. Compute the wind force in the transverse direction acting on the exposed area of the bridge deck and substructure above the lowest bed level. This force is calculated with and without the live loading on the bridge deck. Usually it is sufficient to consider wind on the deck surface area between its soffit and top of the solid parapet or crash barrier, and that of the body of the live load at the given rate as per codal recommendations. Similarly for no live load condition the rate specified by the code is considered. In certain cases the wind force strikes the bridge structure obliquely. In such a case a combination of simultaneously acting wind forces acting in the transverse and longitudinal directions is considered as per codal specifications.

VI. Compute (i) the static equivalent of the horizontal seismic force depending upon the mass of the structure above the earth embedment level which is the maximum scour level in case of hydraulic bridges. The seismic forces are based on the full weight including the submerged portions as long as they are above the embedment level. (ii) The seismic force on a mass can be taken as a certain fraction of its weight, acting through its centroid. This fraction varies depending upon the bridge site being in a severe seismic zone to a non - seismic area as per recommendations of the code. Apart from this a vertical seismic force acting in the upward or downward direction and equal to half of the above mentioned force. This vertical force becomes critical in cases where values of maximum and minimum base pressures are critical. (iii) In case of seismic force acting in the transverse direction, the live load seismic effect is assumed to be acting at 1.2 m above the formation level and that due to the dead load of the deck and superimposed dead load at the deck centroid level. (iv) In case the seismic force is in the direction of the longitudinal axis of the bridge, then the seismic effect due to live load is to be ignored since the braking force is already being considered and any further longitudinal force on the live load will only cause skidding of the wheels.

VII. Estimate the 'active' earth pressure force and moment (at various levels and at the base of foundation) on account of the earthfill retained upto foundation base in case of founding strata being soil or upto rock level in case of rocky founding strata. Passive relief from the front fill is generally to be ignored, but if it is well protected then it may be considered depending upto the actual conditions in situ (but accounting for the negative surcharge angle effect if sloping downwards). In case of wells where the foundation is usually taken down well into permanent and unexecuted soil (e.g. a caisson taken below the maximum scour level), estimate and take into account the net 'passive less active' earth pressure relief (force and moment) as a result of the soil grip.

VIII. Compute the net vertical load, the net horizontal forces in the longitudinal and transverse directions and the net moments about these two directions, at the base of the foundation (at soffit of pile cap in the case of piles) under each critical load combination. Apart from the pile foundations, calculate the base pressures and the safety factors for stability against overturning and sliding and ensure that the requirements are satisfied. In case one of the parametres is not satisfied, then redesign by taking a new set of dimensions.

IX. In case of piled foundations estimate the maximum and minimum axial loads in the piles by the using rivet-group approach. Ensure that no pile is in tension unless tension-piles are provided. Ensure the structural design of the pile-section for the vertical loads and moments combinations on it. Check for the adequacy of soil resistance around an individual pile, the block failure and group action of the pile group.

X. After satisfying the stability requirements of the foundation, calculate the vertical load and moments at various intermediate critical sections (including at the maximum bending moment section in a caisson) in the foundation structure (this includes whole pier/abutment structure), and then structurally design these sections. Also design various other structural components of the foundations structure, taking into account the respective critical load combinations for these components.

XI. Estimate the vertical loads on each bearing under various load combinations, along with the coexisting longitudinal and transverse horizontal forces and rotations. Design the bearings for these effects. Use may be made of the standard manufacturers' catalogues for certain standard bearings.

Types of Piers and their Design Aspects

Open Piers : They are classified as :

1. Multiple bent or multiple column
2. Pile bent
3. Cylindrical piers
4. Trestle piers.

Multiple Bent : They are often used on ground. The section of wall at the bottom may or may not be used. It is needed in overpass work where traffic runs parallel close to the bent to reduce damage to the columns in case of accident. It is lighter and may be more economical than the solid pier; but it requires more formwork and material.

Pile Bent : The pile bent is used for low piers over unstable ground. They consist of reinforced concrete or steel piles driven into the ground, provided with a capping at their top to support the main girders. They are laterally connected by reinforced concrete or steel braces. The pile is used both for a support by driving to resistance and for a column by projecting above ground. The outside piles are usually battered.

Cylindrical Piers : They consist of mild steel or cast iron cylinders which are filled with concrete. When cylinders are used apart and suitable bracing is provided.

Trestle Piers : They are used for temporary work and for timber work. They are made up of RCC or steel vertical, horizontal and diagonal members. The connections in steel trestle may be either riveted or welded. RCC trestle type pier consists of circular columns with a bent cap at the top. In order to avoid moments being transferred from deck to the columns concrete hinges are introduced between the top of the columns and the bent cap. In case of fly overs and elevated roads where trestle piers are used, it is desirable to provide connecting diaphragms between the column.

The following are some of the special piers :

1. Separate piers
2. Abutment piers

3. Cellular piers
4. Framed piers.

Separate piers : They are sometimes used for high bridges. These may be low piers with steel columns above on concrete piers upto the bridge level. In water, they may be combined with a solid shaft below upto flood water level. They may have the advantage of lower cost than solid shafts.

Abutment Piers : In the case of multiple span arch bridges, every third or fourth pier is designed as an abutment to receive the thrust from either side. Such a pier is thicker in section and is known as abutment pier. The object of the abutment pier is listed below :

i. The arches can be turned in sets, and thus heavy expenditure in centreing is saved
ii. In the event of damage, occurring to any portion of the bridge due to floods, the same will probably not extend further than the next abutment.

Cellular Piers: The cellular piers consists of two concentric RCC Shells connected by radial ribs and horizontal bands at suitable intervals. The intermediate space between the shell is filled with some filler material. This will result in saving of cement concrete which is a costly material. This type of piers usually require difficult shuttering and additional labour in placing reinforcements.

Framed Piers : This type of piers result in reduced effective span length for girders on ether side of the centre line of the pier leading to economy in the cost of superstructure. In this type of design two expansion joints at close intervals to 1 to 2 m are provided which results in uneven surface and maintenance problems. Such type of piers if used in rivers subjected to sudden floods near hills and forests would be subjected to floating debris e.g. floating tree trunk etc.

Design of Piers

The following loads and forces are taken into account :

- Dead weight of the Pier
- End reactions due to dead and live loads of superstructure : These reactions will cause an eccentricity when one span only is loaded. Also some end details are so standardised as to dimension that eccentricity which may be caused by difference in span lengths.
- Impact forces due to floating debris and live load for the top 3 m of the pier only.

- Wind pressure acting on the superstructure, moving loads and the pier.
- Centrifugal forces in the event of a road bridge situated on a curve.
- Longitudinal forces arising either due to tractive effort caused through acceleration of the vehicle or braking effect are resulting from the application of the brakes and due to friction at bearings.
- Water pressure
- Earthquake forces in the event of bridge located in seismic zone.
- Buoyancy effect of submerged part of substructure.

The magnitude of the above forces can be determined as per IRC Standard specifications and code of practice section II. Bridge spanning across a tidal river or an estuary, are subjected to an additional hydrodynamic forces due to wave action. The wave motion is essentially an unsteady fluid flow. The fluid particles are subjected to not only a velocity in the horizontal and vertical directions but also to acceleration in these directions. Thus, the fluid particles velocity and acceleration induce a horizontal drag force and an inertial force. The forces developed in the vertical direction are insignificant because of the piers are normally rigidly fixed at the bottom.

Stability of Bridge Piers

Bridge piers are considered stable if they satisfy the following conditions :

1. All the components parts of the pier must be sufficiently strong to resist bending moments and shearing forces.
2. The pier should be stable against sliding downstream and sliding parallel to the axis of the bridge.
3. The pier must not be turned over about the downstream toe and also in the direction of the axis of the bridge.
4. Under the worst combinations of all loads and forces, no tension should be permitted under the foundations, and the maximum pressure should not exceed the safe bearing capacity of the soil.

3

Deconstruction

In the context of physical construction, deconstruction is the selective dismantlement of building components, specifically for re-use, recycling, and waste management. It differs from demolition where a site is cleared of its building by the most expedient means. Deconstruction has also been defined as "construction in reverse". The process of dismantling structures is an ancient activity that has been revived by the growing field of sustainable, green method of building. Buildings, like everything, have a life-cycle. Deconstruction focuses on giving the materials within a building a new life once the building as a whole can no longer continue.

When buildings reach the end of their useful life, they are typically demolished and hauled to landfills. Building implosions or 'wrecking-ball' style demolitions are relatively inexpensive and offer a quick method of clearing sites for new structures. On the other hand, these methods create substantial amounts of waste. Components within old buildings may still be valuable, sometimes more valuable than at the time the building was constructed. Deconstruction is a method of harvesting what is commonly considered "waste" and reclaiming it into useful building material.

Contribution to Sustainability

Deconstruction has strong ties to environmental sustainability. In addition to giving materials a new life cycle, deconstructing buildings helps to lower the need for virgin resources. This in turn leads to energy and emissions reductions from the refining and manufacture of new materials. As deconstruction is often done on a local level, many times on-site, energy and emissions are also saved in the transportation of materials. Deconstruction can potentially support communities by providing local jobs and renovated structures. Deconstruction work typically employs 3-6 workers for every one employed in a comparable demolition job. In addition, solid waste from conventional demolition is diverted from landfills. This is a major benefit because construction and demolition (C&D) waste accounts for approximately 20% of the solid waste stream.

Benefits of Avoiding Wood Waste

In Canada, the CO_2 Neutral Alliance has created a website with resources for regulators and municipalities, developers and contractors, business owners and operators, and individuals and households. Benefits for municipalities include:

- Reducing disposal costs where waste collection, hauling or disposal is supported by the tax base
- Establishing additional revenue streams
- Making existing landfills last longer
- Reducing greenhouse gas emissions caused by the decomposition of wood waste into methane from landfills
- Stimulating local economies with new industries and employment

Improving the local environment and overall sustainability of your community For every three square feet of deconstruction, enough lumber can be salvaged to build one square foot of new construction. At this rate, if deconstruction replaced residential demolition, the United States could generate enough recovered wood to construct 120,000 new affordable homes each year. The deconstruction of a typical 2,000-square-foot (190 m^2) wood frame home can yield 6,000 board feet of reusable lumber. Every year the United States buries about 33 million tons of wood-related construction and demolition debris in landfills. As anaerobic microorganisms decompose this wood, it will release about five million tons of carbon equivalent in the form of methane gas.

Typical Methods of Deconstruction

Deconstruction is commonly separated into two categories; structural and non-structural. Non-structural deconstruction, also known as "soft-stripping", consists of reclaiming non-structural components, appliances, doors, windows, and finish materials. The reuse of these types of materials is commonplace and considered to be a mature market in many locales. Structural deconstruction involves dismantling the structural components of a building. Traditionally this had only been performed to reclaim expensive or rare materials such as used brick, dimension stone, and extinct wood. In antiquity, it was common to raze stone buildings and reuse the stone; it was also common to steal stones from a building that was not being totally demolished: this is the literal meaning of the word *dilapidated.* Used brick and dimension limestone in particular have a long tradition of reuse due to their durability and colour changes over time. Recently, the rise of environmental awareness and sustainable building has made a much wider range of materials worthy of structural deconstruction. Low-end, commonplace materials such as dimensional lumber have become part of this newly emerging market.

The United States military has utilised structural deconstruction in many of its bases. The construction methods of barracks, among other base structures, are usually relatively simple. They typically contained large amounts of lumber and used minimal adhesives and finish-work. In addition, the buildings are often identical, making the process of deconstructing multiple buildings much easier. Many barracks were built during the era prior to WWII, and have aged to the point where they now need to be torn down. Deconstruction was deemed very practical due to the abundance of labour the military has access to and the value of the materials themselves.

Natural disasters, such as hurricanes, floods, tsunamis, and earthquakes often leave a vast amount of usable building materials in their wake. Structures that remain standing are often deconstructed to provide materials for rebuilding the region.

Economic Potential

Deconstruction's economic viability varies from project to project. The amount of time and cost of labour are the main drawbacks. Harvesting materials from a structure can take weeks, whereas demolition may be completed in roughly a day. However, some of the costs, if not all, can be recovered. Reusing the materials in a new on-site structure, selling reclaimed materials, donating materials for income tax write-offs, and avoiding landfill "tipping fees" are all ways in which the cost of deconstruction can be made comparable to demolition.

Reclaiming the materials for a new on-site structure is the most economically and environmentally efficient option. Tipping fees and the costs of new materials are avoided; in addition, the transportation of the materials is non-existent. Selling the used materials or donating them to non-profit organisationsare another effective way of gaining capital. Donations to NPO's such as Habitat for Humanity's ReStore are tax deductible. Many times it is possible to claim the value to be half of what that particular material would cost new. When donating rare or antique components it is sometimes possible to claim a higher value than a comparable, brand-new material.

Value can also be added to new structures that are built by implementing reused materials. The United States Green Building Council's program entitled Leadership in Energy and Environmental Design (LEED) offers seven credits relating to reusing materials. (This accounts for seven out of a maximum sixty-nine credits) These include credits for building-shell reuse, material reuse, and diverting waste from landfills. Building shell-reuse is particularly appropriate for shells made of dimension stone.

Deconstruction is well suited to job training for the construction trades. Taking down a building is an excellent way for a worker to learn how to put a building up. This is vital for the economic recovery of inner-city communities. Unskilled and low-skilled workers can receive on-the-job training in use of basic carpentry tools and techniques, as well as learning teamwork, problem-solving, critical thinking and good work habits.

Process

When choosing to deconstruct a building there are some important aspects that need to be taken into consideration. Developing a list of local contacts that are able to take used materials is an essential first step. These might include commercial architectural salvage businesses, reclamation yards, not-for-profit and social enterprise salvage warehouses, and dismantling contractors. Materials that cannot be salvaged may be recycled on-site or off-site, or taken to landfills. The next step involves identifying which, if any, are hazardous materials. Lead paint and asbestos are two substances in particular that need to be handled extremely cautiously and disposed of properly.

It is common practice, and common sense, to "soft-strip" the structure first; remove all appliances, windows, doors, and other finishing materials. These will account for a large percentage of the marketable components. After the non-structural deconstruction, structural is the next step. It is best to start at the roof and work down to the foundation.

Building components that are dismantled will need to be stored in a secure, dry location. This will protect them from water damage and theft. Once separated from the structure, materials can also be cleaned and/or refinished to increase value. Building an inventory list of the materials at hand will help determine where each item will be sent.

Designing for Deconstruction (DfD)

"The end of the building's useful life generates a stream of used materials that can be reprocessed for new construction. The selection of materials for reuse or recycling should not start at the end of the building's life cycle, it should start at the design stage. Architects and engineers should keep the whole life cycle of the building in mind and select construction materials based on their capacity to be reused or recycled after the building has served its purpose."

An upstream approach to deconstruction can be implemented into buildings during their design process. This is a current trend in sustainable architecture. Often, simple construction methods combined with high-grade, durable materials work best for DfD structures. Separating layers of a building's infrastructure and making them visible can significantly simplify its deconstruction. Making components within systems separable also assists in being able to dismantle materials quickly and efficiently. This can be achieved by using mechanical fasteners such as bolts to connect parts. Allowing physical access to the fasteners is another needed aspect of this design. Also, it is important to use standardized materials and assemble them in a consistent manner throughout the project.

Some conventional construction methods and materials are difficult or impossible to deconstruct and should be avoided when designing for deconstruction. The use of nails and adhesives significantly slows down the deconstruction process and has a tendency to ruin otherwise reusable materials. Avoid hazardous materials altogether as they are detrimental to the natural environment and are non-reusable. Using mixed material grades makes the process of identifying pieces for resale difficult.

Deconstruction is important for more than just the end of a building's life-cycle. Buildings that have been designed with deconstruction in mind are often easier to maintain and adapt to new uses. Saving the shell of a building or adapting interior spaces to meet new needs ensures that new structures have a small environmental impact.

An alternative worth considering is modular building, like the Habitat 67 project in Montreal, Canada. This was a residential structure consisting of separate, functional apartments that could be put together in a variety of ways. As people moved in or out, the units could be reconfigured as desired.

Major Appliance

A major appliance, or domestic appliance, is usually defined as a large machine which accomplishes some routine housekeeping task, which includes purposes such as cooking, or food preservation, whether in a household, institutional, commercial or industrial setting. An appliance is differentiated from a plumbing fixture because it uses an energy input for its operation other than water, generally using electricity or natural gas/propane. An object run by a watermill would also be considered an appliance. The term white goods or whiteware is also used for these items, primarily where British English is spoken, although definitions for the term "white goods" can differ. In the United States, the term *white goods* more commonly refers to linens rather than appliances.

Major appliances have become more technically complex from the control side recently with the introduction of the various Energy Labelling rules across the world. This has meant that the appliances have been forced to become more and more efficient leading to more accurate controllers in order to meet the regulations.

Major appliances are differentiated from small appliances because they are large, difficult to move, and generally fixed in place to some extent. They are often considered fixtures and part of real estate and as such they are often supplied to tenants as part of otherwise unfurnished rental properties. Another frequent characteristic of major appliances is that they may have substantial electricity requirements that necessitate special electrical wiring to supply higher current and/ or voltage than standard electrical outlets can deliver. This limits where they can be placed in a home.

Escalator

An escalator is a moving staircase – a conveyor transport device for carrying people between floors of a building. The device consists of a motor-driven chain of individual, linked steps that move up or down on tracks, allowing the step treads to remain horizontal. Escalators are used around the world to move pedestrian traffic in places where elevators would be impractical. Principal areas of usage include department stores, shopping malls, airports, transit systems, convention centres, hotels, arenas, stadiums and public buildings.

The benefits of escalators are many. They have the capacity to move large numbers of people, and they can be placed in the same physical space as one might install a staircase. They have no waiting interval (except during very heavy traffic), they can be used to guide people toward main exits or special exhibits, and they may be weatherproofed for outdoor use. A non-functioning escalator can function as a normal staircase, whereas many other conveyances become useless when they break down.

Design, Components, and Operation

Operation and Layout: Escalators, like moving walkways, are often powered by constant-speed alternating currentmotors and move at approximately 1–2 feet (0.3–0.6 m) per second.

The typical angle of inclination of an escalator to the horizontal floor level is 30 degrees with a standard rise up to about 60 feet (18 m). Modern escalators have single-piece aluminium or stainless steel steps that move on a system of tracks in a continuous loop.

Figure: *"Multiple parallel" layout*

Escalators have three typical configuration options: parallel (up and down escalators side by side or separated by a distance, seen often in metro stations and multilevel motion picture theaters), crisscross (minimizes space requirements by "stacking" escalators that go in one direction, frequently used in department stores or shopping centres), and multiple parallel (two or more escalators together that travel in one direction next to one or two escalators in the same bank that travel in the other direction).

Figure: *"Crisscross" layout*

Escalators are required to have moving handrails that keep pace with the movement of the steps. This helps riders steady themselves, especially when stepping onto the moving stairs. Occasionally, a handrail will move at a slightly different speed from the steps, causing it to "creep" slowly forward or backward relative to the steps. Riders may find this effect disconcerting and this problem should be fixed.

The direction of escalator movement (up or down) can be permanently set, or be controlled by personnel according to the predominant flow of the crowd, or be controlled automatically. In some setups, direction is controlled automatically by whoever arrives first, whether at the bottom or at the top (the system is programmed so that the direction is not reversed while a passenger is on the escalator).

Design and Layout Considerations

A number of factors affect escalator design, including physical requirements, location, traffic patterns, safety considerations, and aesthetic preferences. Foremost, physical factors like the vertical and horizontal distance to be spanned must be considered.

These factors will determine the length and pitch of the escalator. The building infrastructure must be able to support the heavy components. The escalator should be located where it can be easily seen by the general public. In department stores, customers should be able to view the merchandise easily. Furthermore, up and down escalator traffic should be physically separated and should not lead into confined spaces.

Traffic patterns must also be anticipated. In some buildings, the objective is simply to move people from one floor to another, but in others there may be a more specific requirement, such as funneling visitors towards a main exit or exhibit. The escalators must be designed to carry the required number of passengers.

For example, a single-width escalator travelling at about 1.5 feet (0.5 m) per second can move about 2000 people per hour. The carrying capacity of an escalator system must match the expected peak traffic demand, presuming that passengers ride single file. This is crucial if there are sudden increases in the number of riders. For example, escalators at stations must be designed to cater for the peak traffic flow discharged from a train, without causing excessive bunching at the escalator entrance.

In this regard, escalators help in controlling the flow of people. For example, an escalator to an exit effectively discourages most people from using it as an entrance, and may reduce security concerns. Similarly, escalators often are used as the exit from airport security checkpoints. Such an egress point would generally be staffed to prevent its use as an entrance, as well.

It is preferred that staircases be located adjacent to the escalator if the escalator is the primary means of transport between floors. It may also be necessary to provide an elevator lift near the escalator for wheelchairs and disabled people. Finally, consideration should be given to the aesthetics of the escalator.

Landing Platforms

These two platforms house the curved sections of the tracks, as well as the gears and motors that drive the stairs. The top platform contains the motor assembly and the main drive gear, while the bottom holds the step return idler sprockets. These sections also anchor the ends of the escalator truss. In addition, the platforms contain a floor plate and a combplate.

The floor plate provides a place for the passengers to stand before they step onto the moving stairs. This plate is flush with the finished floor and is either hinged or removable to allow easy access to the machinery below. The combplate is the piece between the stationary floor plate and the moving step. It is so named because its edge has a series of cleats that resemble the teeth of a comb. These teeth mesh with matching cleats on the edges of the steps. This design is necessary to minimize the gap between the stair and the landing, which helps prevent objects from getting caught in the gap.

Truss

The truss is a hollow metal structure that bridges the lower and upper landings. It is composed of two side sections joined together with cross braces across the bottom and just below the top. The ends of the truss are attached to the top and bottom landing platforms via steel or concrete supports. The truss carries all the straight track sections connecting the upper and lower sections.

Tracks

The track system is built into the truss to guide the step chain, which continuously pulls the steps from the bottom platform and back to the top in an endless loop. There are actually two tracks: one for the front wheels of the steps (called the step-wheel track) and one for the back wheels of the steps (called the trailer-wheel track). The relative positions of these tracks cause the steps to form a staircase as they move out from under the combplate. Along the straight section of the truss the tracks are at their maximum distance apart. This configuration forces the back of one step to be at a 90-degree angle relative to the step behind it. This right angle bends the steps into a shape resembling a staircase. At the top and bottom of the escalator, the two tracks converge so that the front and back wheels of the steps are almost in a straight line. This causes the stairs to lay in a flat sheetlike arrangement, one after another, so they can easily travel around the bend in the curved section of track. The tracks carry the steps down along the underside of the truss until they reach the bottom landing, where they pass through another curved section of track before exiting the bottom landing. At this point the tracks separate and the steps once again assume a staircase configuration. This cycle is repeated continually as the steps are pulled from bottom to top and back to the bottom again.

Steps

The steps themselves are solid, one piece, die-cast aluminium or steel. Yellow demarcation lines may be added to clearly indicate their edges. In most escalator models manufactured after 1950, both the riser and the tread of each step is cleated (given a ribbed appearance) with comblike protrusions that mesh with the combplates on the top and bottom platforms and the succeeding steps in the chain. Seeberger- or "step-type" escalators featured flat treads and smooth risers; other escalator models have cleated treads and smooth risers. The steps are linked by a continuous metal chain that forms a closed loop. The front and back edges of the steps are each connected to two wheels. The rear wheels are set further apart to fit into the back track and the front wheels have shorter axles to fit into the narrower front track. As described above, the position of the tracks controls the orientation of the steps.

Handrail

The handrail provides a convenient handhold for passengers while they are riding the escalator. In an escalator, the handrail is pulled along its track by a chain that is connected to the main drive gear by a series of pulleys. It is constructed of four distinct sections. At the centre of the handrail is a "slider", also known as a "glider ply", which is a layer of a cotton or synthetic textile. The purpose of the slider layer is to allow the handrail to move smoothly along its track. The next layer, known as the "tension member", consists of either steel cable or flat steel tape, and provides the handrail with tensile strength and flexibility. On top of tension member are the inner construction components, which are made of chemically treated rubber designed to prevent the layers from separating. Finally, the outer layer—the only part that passengers actually see—is the cover, which is a blend of synthetic polymers and rubber. This cover is designed to resist degradation from environmental conditions, mechanical wear and tear, and human vandalism.

In the factory, handrails are constructed by feeding rubber through a computer-controlled extrusion machine to produce layers of the required size and type in order to match specific orders. The component layers of fabric, rubber, and steel are shaped by skilled workers before being fed into the presses, where they are fused together.

In the mid-twentieth century, some handrail designs consisted of a rubber bellows, with rings of smooth metal cladding called "bracelets" placed between each coil. This gave the handrail a rigid yet flexible feel. Additionally, each bellows section was no more than a few feet long, so if part of the handrail was damaged, only the bad segment needed to be replaced. These forms of handrail have largely been replaced with conventional fabric-and-rubber railings.

Safety

Safety is also major concern in escalator design. In India where women wear saris, there are heavy chances of getting the *pallu* entangled in the escalator. As a result, a special *sari guard* is built into most escalators. There is a risk of feet injuries for children wearing footwear such as Crocs and flip-flops that might get caught in escalator mechanisms. This was due to the softness of the shoe's material combined with the smaller size of children's feet.

Fire protection of an escalator floor opening may be provided by adding automatic sprinklers or fireproof shutters to the opening, or by installing the escalator in an enclosed fire-protected hall. To limit the danger of overheating, ventilation for the spaces that contain the motors and gears must be provided.

Accidents and Litigation

Accidents: There have been reports of people falling off a moving escalator or getting their shoe stuck in part of the escalator; shoe laces are a hazard when loose. Some accidents are caused by improper or unsafe use such as riding the hand rails or by escalator spinning. A few fatal accidents are:

- Eight people died and 30 more were injured on February 17, 1982, when an escalator collapsed on the Moscow Metro. Wrongly set up service brakes were later blamed for the accident.
- 31 people died after a fire, begun in the undercarriage of an MH-type Otis escalator, exploded into the ticketing hall at King's Cross St. Pancras station in 1987.
- On December 13, 1999, 8-year-old Jyotsna Jethani was killed at New Delhi's international airport. Jethani fell into a gaping hole that resulted from improper maintenance.
- On June 15, 2002, Andrea Albright, a 24-year-old J.C. Penney employee in Columbia, Maryland, was injured while riding the store's escalator, and died from her injuries 10 days later.
- On July 2, 2003, 30 people were injured when an escalator at Coors Field in Denver malfunctioned, and one victim eventually had to have a leg amputated. The accident is believed to have

been caused by a combination of overcrowding and an improperly installed safety device.

- On December 31, 2004, escalators at the Taipei City Hall Station kept moving commuters onto the overcrowded island platform. A woman whose hair got caught in the escalator received 20 stitches to the scalp.
- Francisco Portillo died after being strangled when his sweatshirt got caught in an escalator at the Porter SquareMBTA station in Cambridge, Massachusetts on February 21, 2005. He was allegedly drunk at the time.
- On September 13, 2008, an 11-year old boy died after falling off an escalator in Lyngdal, Norway. On April 20, 2009, a teenage boy died after getting very serious skull injuries after falling off an escalator in Falun, Sweden. On June 26, 2009, a man died after falling off an escalator in Helsingborg, Sweden. All three were riding the handrail.
- On July 6, 2011, a 13-year old boy was killed and several people were injured when they were thrown off an escalator that suddenly changed direction in a Beijing subway station. Officials identified the manufacturers of the escalator as a possible reason for the accident. The escalator was built by Otis and was still in its guarantee period.
- On August 3, 2011, 35 year old deputy stationmaster Kuo Cheng-chieh was killed after being ingested into an escalator. He was carrying a long board used for flood defences on an upward-moving escalator at Dingxi Station in New Taipei City, Taiwan. The board became wedged between the ceiling and the step Kuo was standing on, the step collapsed under the load, and the victim fell into the machinery. The event didn't activate any safety mechanisms and the escalator continued to run.
- On August 30, 2012, a man was killed when he fell 5 stories off an escalator rail at Reliant Stadium during a Texans game.
- On January 7, 2013, at least 6 people were injured when the crowded escalator at the Exchange Place PATH station suddenly reversed direction.

Lessons of the King's Cross Fire

The King's Cross fire of 1987 illustrated the demanding nature of escalator upkeep and the devices' propensity to collect "fluff" when not properly maintained. In the official inquiry that followed, the Fennell Report, it was determined that the fire started slowly, smoldered virtually undetected for a time, then exploded into the ticketing hall above in a phenomenon known as the "trench effect." In the escalators' undercarriage, approximately 8,800 kilograms (19,400 lb) of accumulated detritus acted as a wick to a neglected buildup of interior lubricants; wood veneers, paper and plastic advertisements, solvent-based paint, plywood in the ticket hall, and melamine combustion added to the impact of the calamity.

As a result of the report, older wooden escalators were removed from service in the London Underground. Additionally, sections of the London Underground that were actually below ground were made nonsmoking; eventually the whole system became a smoke-free zone.

Litigation

In the 1930s, at least one suit was filed against a department store, alleging that its escalators posed an attractive nuisance, responsible for a child's injury. These cases were almost always dismissed. Moreover, continual updating of escalator safety codes facilitated increased levels of consumer safety as well as a reduction in court cases.

Legislation and Escalators

United States: Despite their considerable scope, two Congressional Acts, the Rehabilitation Act of 1973 and the Americans with Disabilities Act of 1990 (ADA), did not directly affect escalators or their public installations. Since Section 504 of the Rehabilitation Act included public transportation systems, for a few years, the United States Department of Transportation considered designs to retrofit existing escalators for wheelchair access. Nonetheless, Foster-Miller Associates' 1980 plan, *Escalator Modification for the Handicapped* was ultimately ignored in favour of increased elevator installations in subway systems. Likewise, the ADA provided more accessibility options, but expressly excluded escalators as "accessible means of egress," advocating neither their removal nor retention in public structures.

Codes and Regulation

In the United States and Canada, new escalators must abide by ASMEA17.1 standards, and old/historic escalators must conform to the safety guidelines of ASMEA17.3. In Europe, the escalator safety code is EN115.

Key Safety Features

To enhance passenger safety, newer models of escalators are equipped with one or more of the following safety implementations, as per ASMEA17.1 code:

- Antislide devices: Raised circular objects that often stud the escalator balustrade. Sometimes informally called "hockey pucks" due to their appearance, their purpose is to prevent objects (and people) from precipitously sliding down the otherwise smooth metallic surface.
- Combplate impact switches: Stop the escalator if a foreign object gets caught between the steps and the combplate on either end.
- Deflector brush: A long continuous brush made of stiff bristles running up the sides of the escalator just above the step level. This helps deflect garments, shoes, and other items away from the gap between the moving steps and the skirt board.
- Emergency stop button: At each end of the escalator (in some models, also on the balustrade), a large red button can be pressed to stop the device in the event of an emergency. Typically, an alarmed transparent plastic guardplate covers the button; restarting requires turning a key.
- Extended balustrades: Allows riders to grasp the handrail before setting foot on an escalator, to ease customer comfort and stability/equilibrium. (The effect is similar to the flat steps described below.)
- Flat steps: Like a moving walkway, the first two or three steps at either end of the escalator are flat. This gives the passenger extra time to orient him/herself when boarding, and more time to maintain balance when exiting. Longer escalators often have four or more flat steps.
- Handrail inlet switches: Sensors located at the bottom and top of the unit that guard the handrail termini. If something gets caught in these locations, a hard fault is generated in the controller, and the escalator shuts down automatically.
- Handrail speed sensors: These sensors are usually optical, and monitor how fast the handrail moves. If the sensor notices a speed difference between the handrail and the steps, it sounds an alarm, pauses, and then automatically stops the escalator. In these situations, the escalator must be serviced by authorized personnel before returning to an operable state.

- Missing step detectors: Depending on the manufacturer and model, this sensor is either optical or physical. When a missing step is detected, the escalator automatically shuts down.
- Raised step edges: In some models, a difference in tread height is utilised to keep passengers' feet from the skirt board.
- Safety instructions: A sign, typically posted on both escalator newels at the entrance landing platform. In some situations, safety precautions are posted on walls near the escalator, included on freestanding signs, or—as in some models—printed on the riser surface itself.
- Sensor switch: In automatic-start/stop escalators, this sensor automatically engages the escalator motion when a rider is detected on the first step of the entrance landing platform, and stops the escalator when there are no riders on the unit.
- Step demarcation lights: Either fluorescent or LED lights (traditionally green in colour) located inside the truss. The illumination between the steps improves the passengers' awareness of the step divisions.
- Step demarcation lines: In order to clearly delineate the edges of each individual step, manufacturers offer steps trimmed in yellow, either painted or with plastic inserts.

History

Inventors and manufacturers:

Nathan Ames: Nathan Ames, a patent solicitor from Saugus, Massachusetts, is credited with patenting the first "escalator" in 1859, despite the fact that no working model of his design was ever built. His invention, the "revolving stairs", is largely speculative and the patent specifications indicate that he had no preference for materials or potential use (he noted that steps could be upholstered or made of wood, and suggested that the units might benefit the infirm within a household use), though the mechanization was suggested to run either by manual or hydraulic power.

Leamon Souder

In 1889, Leamon Souder successfully patented the "stairway", an escalator-type device that featured a "series of steps and links jointed to each other". No model was ever built. This was the first of at least four escalator-style patents issued to Souder, including two for spiral designs (U. S. Patent Nos. 723,325 and 792,623).

Jesse Wilford Reno, George A. Wheeler, and Charles Seeberger

In 1892, Jesse W. Reno patented the "Endless Conveyor or Elevator." A few months after Reno's patent was approved, George A. Wheeler patented his ideas for a more recognisable moving staircase, though it was never built. Wheeler's patents were bought by Charles Seeberger; some features of Wheeler's designs were incorporated in Seeberger's prototype built by the Otis Elevator Company in 1899.

Reno, a graduate of Lehigh University, produced the first working escalator (he actually called it the "inclined elevator") and installed it alongside the Old Iron Pier at Coney Island, New York City in 1896. This particular device was little more than an inclined belt with cast-iron slats or cleats on the surface for traction, and travelled along a 25° incline. A few months later, the same prototype was used for a month long trial period on the Manhattan side of the Brooklyn Bridge. Reno eventually joined forces with Otis, and retired once his patents were purchased outright. Some Reno-type escalators were still being used in the Boston subway until construction for the Big Dig precipitated their removal. The Smithsonian Institution considered re-assembling one of these historic units from 1914 in their collection of Americana, but "logistics and reassembly costs won out over nostalgia", and the project was discarded.

Around May 1895, Charles Seeberger began drawings on a form of escalator similar to those patented by Wheeler in 1892. This device actually consisted of flat, moving stairs, not unlike the escalators of today, except for one important detail: the step surface was smooth, with no comb effect to safely guide the rider's feet off at the ends. Instead, the passenger had to step off sideways. To facilitate this, at the top or bottom of the escalator the steps continued moving horizontally beyond the end of the handrail (like a miniature moving sidewalk) until they disappeared under a triangular "divider" which guided the passenger to either side. Seeberger teamed with Otis in 1899, and together they produced the first commercial escalator which won the first prize at the Paris 1900 *Exposition Universelle* in France. Also on display at the *Exposition* were Reno's inclined elevator, a similar model by James M. Dodge and the Link Belt Machinery Co., and two different devices by French manufacturers Hallé and Piat.

Early European Manufacturers: Hallé, Hocquardt, and Piat

Piat installed its "stepless" escalator in Harrods Knightsbridge store on Wednesday, November 16, 1898, though the company relinquished its patent rights to the department store. Noted by Bill

Lancaster in *The Department Store: a Social History*, "customers unnerved by the experience were revived by shopmen dispensing free smelling salts and cognac." The Harrods unit was a continuous leather belt made of "224 pieces . . . strongly linked together travelling in an upward direction," and was the first "moving staircase" in England.

Hocquardt received European patent rights for the *Fahrtreppe* in 1906. After the *Exposition*, Hallé continued to sell its escalator device in Europe, but was eventually eclipsed in sales by other major manufacturers.

Major Competitors and Product Nomenclature

In the first half of the twentieth century, several manufacturers developed their own escalator products, though they had to market their devices under different names, due to Otis' hold on the trademark rights to the word "escalator." New York-based Peelle Company called their models the *Motorstair*, and Westinghouse called their model an *Electric Stairway*. The Toledo-based Haughton Elevator company referred to their product as simply *Moving Stairs*.

Manufacturing Mergers and Buyouts: the Playing Field Narrows

Kone and Schindler introduced their first escalator models several decades after the Otis Elevator Co., but grew to dominance in the field over time. Today, they, Mitsubishi, and ThyssenKrupp are Otis' primary rivals. Schindler now stands as the largest maker of escalators and second largest maker of elevators in the world, though their first escalator installation did not occur until 1936. In 1979, the company entered the United States market by purchasing Haughton Elevator; nine years later, Schindler assumed control of the North American escalator/elevator operations of Westinghouse.

Kone expanded internationally by acquisition in the 1970s, buying out Swedish elevator manufacturer Asea-Graham, and purchasing other minor French, German, and Austrian elevator makers before assuming control of Westinghouse's European elevator business. As the last "big four" manufacturers held on to the escalator market, KONE first acquired Montgomery Elevator Company, then took control of Germany's Orenstein & Koppel *Rolltreppen*.

Model Development and Design Types

"Cleat-Type" Escalators: Jesse Reno's escalators did not resemble modern escalators too closely. Passengers' feet tilted upward at an angle, and the treads consisted of cleated metal (initially) or wood (later models). Reno worked on his own for several years, gaining success with installations from Toronto to Cape Town, South Africa. Similar units of the day by other manufacturers resembled conveyor belts more than moving staircases. For a time, Otis Elevator sold Reno's escalators as their own "cleat-type" escalators.

"Step-Type" Escalators

Seeberger's model, bought by Otis, clearly became the first "step-type" escalator, so called after its visual likeness to steps on a regular staircase. The company later combined the best aspects of both inventions (guiding slats and flat steps) and in 1921 produced an escalator similar to the type used today: they called it the "L-type" escalator. It was succeeded by the "M-type", the "O-type", and current models by Otis such as the "NCE-type" escalator.

Spiral Escalators

Jese Reno, in addition to being known for the first "practical" escalator in public use, also bears the unique distinction of designing the very first escalators installed in any underground subway system – a single spiral escalator at Holloway Road tube station in London in 1906. The experimental device never saw public use, and was forgotten for several decades. The remains of this are now in the London Transport Museum's depot in Acton. Also the first fully operational spiral escalator, Reno's design was nonetheless only one in a series of several similar proposed contraptions. Souder patented two spiral designs, Wheeler drafted spiral stairway plans in 1905, Seeberger devised at least two different spiral units between 1906 and 1911 (including an unrealised arrangement for the London Underground), and Gilbert Luna obtained West German, Japanese, and United States patents for his version of a spiral escalator by 1973. When interviewed for the *Los Angeles Times* that year, Luna was in the process of soliciting "major firms" for acquisition of his patents and company, but statistics are unclear on the outcome of his endeavours in that regard.

The Mitsubishi Electric Corporation was most successful in its development of "spiral" (more "curve" than true spiral) escalators, and has sold them exclusively since the mid-1980s. The world's first "practical" spiral escalator—a Mitsubishi model—was installed in Osaka, Japan, in 1985. In use, a major planning advantage presented by spiral escalators is that they take up much less horizontal floor space than traditional units, which frequently house large machine rooms underneath the truss.

Helixator, an experimental spiral escalator design which currently exists as a prototype scale model, could further reduce the floor space. The design has several innovations that offer the possibility of a continuous helix. Driven by a linear motor instead of a chain system, the Helixator spreads the force evenly along the escalator path, avoiding excessive force on the top chain links and removing geometry, length and height limits.

Etymology

Several authors and historians have contributed their own differing interpretations of the source of the word "escalator", and some degree of misinformation has heretofore proliferated on the Internet. For reference, contradictory citations by seven separate individuals, including the Otis Elevator Company itself, are provided below.

Name Development and Original Intentions

Charles Seeberger created the word "escalator" in 1900, to coincide with his device's debut at the *Exposition Universelle.* According to his own account, in 1895, his legal counsel advised him to name his new invention, and he then set out to devise a title for it on his own. As evidenced in Seeberger's own handwritten documents, archived at the Otis Elevator Company headquarters in Farmington, Connecticut, the inventor consulted "a Latin lexicon" and "adopted as the root of the new word, 'Scala'; as a prefix, 'E' and as a suffix, 'Tor.'" His own rough translation of the word thus created was "means of traversing from", and he intended for the word to be pronounced, "es' ¹kæl' Y' tYr" (es-CAL-a-tor).

"Escalator" was not a combination of other French or Greek words, and was never a derivative of "elevator" in the original sense, which means "one who raises up, a deliverer" in Latin. Similarly, the root word "*scala*" does not mean "a flight of steps", but is defined by Lewis and Short's *A Latin Dictionary* as the singular form of the plural noun "*scalae*", which denotes any of the following: "a flight of steps or stairs, a staircase; a ladder, [or] a scaling-ladder."

The alleged intended capitalization of "escalator" is likewise a topic of debate. Seeberger's trademark application lists the word not only with the "E" but also with all of the letters capitalized (in two different instances), and he specifies that, "any other form and character of type may be employed . . . without altering in any essential manner the character of [the] trade-mark." That his initial specifications are ostensibly inconsistent, and since Otis Elevator Co. advertisements so frequently capitalized all of the letters in the word, suppositions about the "capital 'e'" are difficult to formulate.

Derivatives of 'Escalator'

The verb "escalate" originated in 1922, and has two uses, the primary: "to climb or reach by means of an escalator" or "to travel on an escalator", and the secondary: "to increase or develop by successive stages; *spec.* to develop from 'conventional' warfare into nuclear warfare." The latter definition was first printed in the *Manchester Guardian* in 1959, but grew to prominent use during the late 1960s and early 1970s.

Loss of Trademark Rights

In 1950, the landmark case *Haughton Elevator Co. v. Seeberger* precipitated the end of Otis's reign over exclusive use of the word "escalator", and simultaneously created a cautionary study for companies and individuals interested in trademark retention. Confirming the contention of the Examiner of Trademark Interferences, Assistant Commissioner of Patents Murphy's decision rejected the Otis Elevator Company's appeal to keep their trademark intact, and noted that "the term 'escalator' is recognised by the general public as the name for a moving stairway and not the source thereof", observing that the Otis Elevator Co. had "used the term as a generic descriptive term...in a number of patents which [had] been issued to them and...in their advertising matter." All trademark protections were removed from the word "escalator", the term was officially genericized, and it fell into the public domain.

Primary Uses and Application

Department Stores/Shopping: As noted above, a few escalator types were installed in major department stores (including Harrods) before the *Expo*. Escalators proved instrumental in the layout and design of shopping venues in the twentieth century. By 1898, the first of Reno's "inclined elevators" were incorporated into the Bloomingdale Bros. store at Third Avenue and 59th Street. This was the first retail application of the devices in the US, and no small coincidence, considering that Reno's primary financier was Lyman Bloomingdale, co-owner of the department store with brother Joseph Bloomingdale.

Public Transportation

The first "standard" escalator installed on the London Underground was a Seeberger model at Earls Court. Noted above, London's

Underground installed a rare spiral escalator designed by Reno, William Henry Aston and Scott Kietzman for the Holloway Road Underground station in 1906; it was run for a short time but was taken out of service the same day it debuted. The older lines of the London Underground had many escalators with wooden treads (ca. 1930s) until they were rapidly replaced following the King's Cross fire, noted above.

Other Applications

Factories and Other Industrial Production Environments: In 1905, the American Woolen Company's Wood Mill in Lawrence, Massachusetts (then "the largest single worsted mill in the world") utilised Otis' Seeberger-type "reversible" escalators to carry its workers between floors four times a day. The machines did not run all day: rather, escalators ran solely to transport employees to/from midday meals and in/out of the mill. In its advertising, Otis Elevator Company hailed this unconventional use for its unique benefits to both workers and owners: "The profitable and practicable feature of the Escalator, from the viewpoint of the owner, is the increased efficiency of each operator due to the elimination of stair climbing."

Military Use

In San Francisco, an escalator at Hunters Point Naval Shipyard was used to convey personnel between the first and third floors. At the time of its construction in 1948, it was touted thus: "[it has the] highest lift of any industrial building in the world. It rises 42 feet (12.8 metres)."

Escalators were also utilised on aircraft carriers such as the USS *Hornet* (CV-12), to transport pilots from "ready rooms" to the flight deck.

Extant Historic Escalator Models

A few notable examples of historic escalators still in operation are:

Australia

- Town Hall Railway Station, Sydney, Australia
- Wynyard Railway Station, Sydney, Australia

Europe

- St. Anna Pedestrian Tunnel underneath the Schelde in Antwerp, Belgium. This tunnel was opened in 1933.
- Tyne Cyclist and Pedestrian Tunnel, Tyne and Wear, England

These escalators, manufactured by Waygood Otis in 1951, were "believed to be the longest single lift escalators in the world", at the time of installation. Presumably the first escalators in Britain designed specifically for cyclists, they were also the longest in the United Kingdom. At most, they may be the longest extant wooden escalators in the world, though they are no longer in working order.

- Baumanskayametro station in Moscow, Russia contains the oldest (1944) escalator of tonnel type still in use, with length of 65 m (213 ft) and vertical rise of 32.5 m (107 ft).
- Greenford Station in Greenford, England contains the oldest (1947) escalator still in use on the London Underground service. This was the first above ground station to install an escalator to its high levels and one of the few (if not the only) wooden escalators not to be replaced after the fire at Kings Cross station.

North America

- Macy's Herald Square department store, Otis L-type units with wood treads and replacement metal treads, New York, New York (only in escalators heading upwards from the first to the seventh stories)
- Kaufmann's department store (now Macy's), two 16-inch (406 mm) Otis L-type units with original floorplates, several 40-inch (1,016 mm) Otis escalators ca. 1950s, Pittsburgh, Pennsylvania
- Westfield San Francisco Centre (formerly The Emporium), chrome-and-glass escalator by Eleanor LeMaire for Otis, San Francisco, California
- The Land Pavilion-Epcot, Glass escalator by Otis. Has stairs next to them. Not single file but is separated
- Tokyo Sesame Place-Glass escalator with Big Bird's mouth as the entrance. First escalator in a theme park (now closed)

Escalators: Superlatives

Longest systems:

- Central-Mid-Levels escalator: in Hong Kong, tens of thousands of commuters travel each work day between Central, the central business district, and the Mid-levels, a residential district hundreds of feet uphill, using this long distance system of escalators and moving walkways. It is the world's longest outdoor escalator *system* (not a single escalator span), at a total length of 2,600 feet (790 m). It goes only one way at a time; the direction reverses depending on rush hour traffic direction.

- In December 2011, a network of six escalators of 1,260 feet (380 m) length, equivalent to 28 stories high, was opened in Medellin, Colombia, offering the 12,000 residents of Comuna 13 a six-minute ride to the city centre compared to their previous 35-minute climb on foot.
- Cascade, Yerevan: an escalator system of 774 ft (236 m) length and 387 ft (118 m) height.
- Ocean Park, Hong Kong: a long escalator system connecting two parts of the Park, with an overall length of 730 feet (220 m).

Longest Individual Escalators

World:

- The longest escalators in the world are installed in deep underground stations of the Saint Petersburg Metro. Three stations have escalators up to 132 m (433 ft) long and 66 m (217 ft) high: Ploshchad Lenina, Chernyshevskaya, and Admiralteyskaya.
- The longest *freestanding* (supported only at the ends) escalator in the world is inside CNN Centre's atrium in Atlanta, Georgia. It rises 8 stories and is 205 feet (62 m) long. Originally built as the entrance to the amusement park *The World of Sid and Marty Krofft*, the escalator is now used for CNN studio tours.

Asia:

- The longest escalator in Bangkok, Thailand and Southeast Asia is in Bangkok Metro's Si Lom Station. It connects the concourse level with platform 1 which in turn connects to Hua Lam Phong. It is 43 m (141 ft) in length and 21.5 m (71 ft) in depth.

Australia:

- The longest set of single-span uninterrupted escalators in the southern hemisphere is at Parliament underground railway station in Melbourne.

Europe:

Several metro or subway systems in Central and Eastern Europe feature very long escalators.

- The Kiev MetroKreschatik station's lower-level second exit escalator (a type ËÒ-2, circa 1965), lifts riders 65.7 metres (216 ft), or 743 steps, up a 131.4-metre (431 ft) – long incline.
- In the Park Pobedy station of the Moscow Metro, the escalators are 126.8 m (416 ft) or 740 steps long, and 63.4 m (208 ft) high. It takes three minutes to transit.
- The longest escalator in Prague, and in European Union, is at the NámìstíMíru station at 285 feet (87 m) long and 142.5 feet (43.4 m) high.
- The longest escalator of a European shopping mall is at MyZeil, Frankfurt, Germany, with a length of 150 feet (46 m).
- The tallest escalator on the London Underground system is at Angel station with a length of 200 feet (61 m), and a vertical rise of 90 feet (27 m).
- The longest wooden escalators in the United Kingdom are at the Tyne Tunnel, with a length of 197 feet (60 m).
- The longest escalators in Western Europe are in the Stockholm Metro, at Västraskogen with a length of 220 feet (67 m) and in Helsinki Metro, at Kamppi station with a length of 210 feet (64 m).
- The largest "single truss escalator" is in the Bentall Centre in Kingston upon Thames in Greater London, UK. It connects the ground floor with the second floor with only top and bottom supports.

North and South America

- The longest set of single-span uninterrupted escalators in the Western Hemisphere is at the Wheaton station of the Washington Metro system. They are 230 feet (70 m) long with a vertical rise of 115 feet (35 m), and take what is variously described as 2 minutes and 45 seconds or nearly three-and-a-half minutes, to ascend or descend without walking.

Shortest Examples

According to Guinness, the shortest escalator in the world is in the Okadaya Mores shopping mall in Kawasaki, Japan. Its vertical rise is only 32.8 inches (83 cm).

Electrical Wiring

Electrical wiring in general refers to insulated conductors used to carry electricity, and associated devices. This article describes general aspects of electrical wiring as used to provide power in buildings and structures, commonly referred to as building wiring. This article is intended to describe common features of electrical wiring that may apply worldwide. For information regarding specific national electrical codes, refer to the articles mentioned in the next section. Separate articles cover long-distance electric power transmission and electric power distribution.

Wiring Safety Codes

Wiring safety codes are intended to protect people and property from electrical shock and fire hazards. Regulations may be established by city, county, provincial/state or national legislation, usually by adopting a model code (with or without local amendments) produced by a technical standards-setting organisation, or by a national standard electrical code.

Electrical codes arose in the 1880s with the commercial introduction of electrical power. Many conflicting standards existed for the selection of wire sizes and other design rules for electrical installations.

The first electrical codes in the United States originated in New York in 1881 to regulate installations of electric lighting. Since 1897 the US National Fire Protection Association, a private non-profit association formed by insurance companies, has published the *National Electrical Code* (NEC). States, counties or cities often include the NEC in their local building codes by reference along with local differences. The NEC is modified every three years. It is a consensus code considering suggestions from interested parties. The proposals are studied by committees of engineers, tradesmen, manufacturer representatives, fire fighters, and other invitees.

Since 1927, the Canadian Standards Association (CSA) has produced the Canadian *Safety Standard for Electrical Installations*, which is the basis for provincial electrical codes. The CSA also produces the Canadian Electrical Code, the 2006 edition of which references IEC 60364 (*Electrical Installations for Buildings*) and states that the code addresses the fundamental principles of electrical protection in Section 131. The Canadian code reprints Chapter 13 of IEC 60364, but there are no numerical criteria listed in that chapter to assess the adequacy of any electrical installation.

Although the US and Canadian national standards deal with the same physical phenomena and broadly similar objectives, they differ occasionally in technical detail. As part of the North American Free

Trade Agreement (NAFTA) program, US and Canadian standards are slowly converging toward each other, in a process known as harmonization. In European countries, an attempt has been made to harmonize national wiring standards in an IEC standard, IEC 60364 *Electrical Installations for Buildings*. Hence national standards follow an identical system of sections and chapters. However, this standard is not written in such language that it can readily be adopted as a national wiring code. Neither is it designed for field use by electrical tradesmen and inspectors for testing compliance with national wiring standards. By contrast, national codes, such as the NEC or CSA C22.1, generally exemplify the common objectives of IEC 60364, but provide specific rules in a form that allows for guidance of those installing and inspecting electrical systems.

In Germany, DKE (the German Commission for Electrical, Electronic and Information Technologies of DIN and VDE) is the organisation responsible for the promulgation of electrical standards and safety specifications. DIN VDE 0100 is the German wiring regulations document harmonised with IEC 60364. In the United Kingdom, wiring installations are regulated by the Institution of Engineering and Technology *Requirements for Electrical Installations: IEE Wiring Regulations, BS 7671: 2008,* which are harmonised with IEC 60364. The 17th edition (issued in January 2008) includes new sections for microgeneration and solar photovoltaic systems. The first edition was published in 1882.

In Australia and New Zealand, the AS/NZS 3000 standard, commonly known as the "wiring rules", specifies requirements for the selection and installation of electrical equipment, and the design and testing of such installations. The standard is mandatory in both New Zealand and Australia; therefore, all electrical work covered by the standard must comply. The international standard wire sizes are given in the IEC 60228 standard of the International Electrotechnical Commission. In North America, the American Wire Gauge standard for wire sizes is used.

Colour Code

To enable wires to be easily and safely identified, all common wiring safety codes mandate a colour scheme for the insulation on power conductors. In a typical electrical code, some colour-coding is mandatory, while some may be optional. Many local rules and exceptions exist. Older installations vary in colour codes, and colours may shift with insulation exposure to heat, light, and aging.

Many electrical codes now recognise (or even require) the use of wire covered with green insulation, additionally marked with a prominent yellow stripe, for safety grounding (earthing) connections. This growing international standard was adopted for its distinctive appearance, to reduce the likelihood of dangerous confusion of safety grounding wires with other electrical functions, especially by persons affected by red-green colour blindness.

Wiring Methods

Materials for wiring interior electrical systems in buildings vary depending on:

- Intended use and amount of power demand on the circuit
- Type of occupancy and size of the building
- National and local regulations
- Environment in which the wiring must operate.

Wiring systems in a single family home or duplex, for example, are simple, with relatively low power requirements, infrequent changes to the building structure and layout, usually with dry, moderate temperature, and non-corrosive environmental conditions. In a light commercial environment, more frequent wiring changes can be expected, large apparatus may be installed, and special conditions of heat or moisture may apply. Heavy industries have more demanding wiring requirements, such as very large currents and higher voltages, frequent changes of equipment layout, corrosive, or wet or explosive atmospheres. In facilities that handle flammable gases or liquids, special rules may govern the installation and wiring of electrical equipment in hazardous areas. Wires and cables are rated by the circuit voltage, temperature rating, and environmental conditions (moisture, sunlight, oil, chemicals) in which they can be used. A wire or cable has a voltage (to neutral) rating, and a maximum conductor surface temperature rating. The amount of current a cable or wire can safely carry depends on the installation conditions.

Figure: *Installing electrical wiring by cutting into the bricks of the building*

Early Wiring Methods

The first interior power wiring systems used conductors that were bare or covered with cloth, which were secured by staples to the framing of the building or on running boards. Where conductors went through walls, they were protected with cloth tape. Splices were done similarly to telegraph connections, and soldered for security. Underground conductors were insulated with wrappings of cloth tape soaked in pitch, and laid in wooden troughs which were then buried. Such wiring systems were unsatisfactory because of the danger of electrocution and fire, plus the high labour cost for such installations.

Knob and Tube

The earliest standardized method of wiring in buildings, in common use in North America from about 1880 to the 1930s, was *knob and tube* (K&T) wiring: single conductors were run through cavities between the structural members in walls and ceilings, with ceramic tubes forming protective channels through joists and ceramic knobs attached to the structural members to provide air between the wire and the lumber and to support the wires. Since air was free to circulate over the wires, smaller conductors could be used than required in cables. By arranging wires on opposite sides of building structural members, some protection was afforded against short-circuits that can be caused by driving a nail into both conductors simultaneously.

By the 1940s, the labour cost of installing two conductors rather than one cable resulted in a decline in new knob-and-tube installations. However, the US code still allows new K&T wiring installations in special situations (some rural and industrial applications).

Metal-Sheathed Wires

In the United Kingdom, an early form of insulated cable, introduced in 1896, consisted of two impregnated-paper-insulated conductors in an overall lead sheath. Joints were soldered, and special fittings were used for lamp holders and switches. These cables were similar to underground telegraph and telephone cables of the time. Paper-insulated cables proved unsuitable for interior wiring installations because very careful workmanship was required on the lead sheaths to ensure moisture did not affect the insulation.

A system later invented in the UK in 1908 employed vulcanized-rubber insulated wire enclosed in a strip metal sheath. The metal sheath was bonded to each metal wiring device to ensure earthing continuity.

A system developed in Germany called "Kuhlo wire" used one, two, or three rubber-insulated wires in a brass or lead-coated iron sheet tube, with a crimped seam. The enclosure could also be used as a return conductor. Kuhlo wire could be run exposed on surfaces and painted, or embedded in plaster. Special outlet and junction boxes were made for lamps and switches, made either of porcelain or sheet steel. The crimped seam was not considered as watertight as the

Stannos wire used in England, which had a soldered sheath. A somewhat similar system called "concentric wiring" was introduced in the United States around 1905. In this system, an insulated electrical wire was wrapped with copper tape which was then soldered, forming the grounded (return) conductor of the wiring system. The bare metal sheath, at earth potential, was considered safe to touch.

While companies such as General Electric manufactured fittings for the system, and a few buildings were wired with it, it was never adopted into the US National Electrical Code. Drawbacks of the system were that special fittings were required, and that any defect in the connection of the sheath would result in the sheath becoming energized.

Other Historical Wiring Methods

Other methods of securing wiring that are now obsolete include:

- Re-use of existing gas pipes when converting gas light installations to electric lighting. Insulated conductors were pulled through the pipes that had formerly supplied the gas lamps. Although used occasionally, this method risked insulation damage from sharp edges inside the pipe at each joint.
- Wood mouldings with grooves cut for single conductor wires, covered by a wooden cap strip. These were prohibited in North American electrical codes by 1928. Wooden moulding was also used to some degree in England, but was never permitted by German and Austrian rules.
- A system of flexible twin cords supported by glass or porcelain buttons was used near the turn of the 20th century in Europe, but was soon replaced by other methods.
- During the first years of the 20th century, various patented forms of wiring system such as Bergman and Peschel tubing were used to protect wiring; these used very thin fibre tubes, or metal tubes which were also used as return conductors.
- In Austria, wires were concealed by embedding a rubber tube in a groove in the wall, plastering over it, then removing the tube and pulling wires through the cavity.

Metal moulding systems, with a flattened oval section consisting of a base strip and a snap-on cap channel, were more costly than open wiring or wooden moulding, but could be easily run on wall surfaces. Similar surface mounted raceway wiring systems are still available today.

Cables

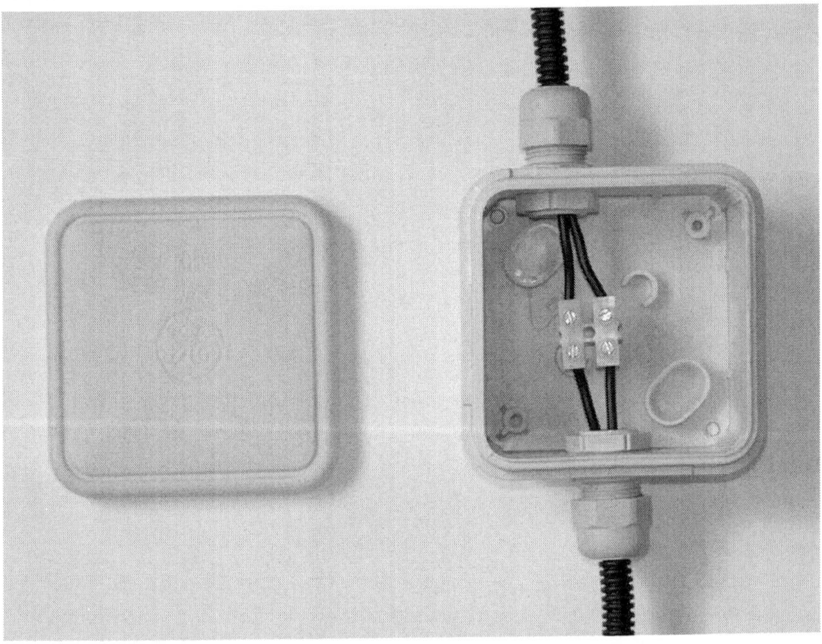

Figure: *Wiring for extremely wet conditions*

Armoured cables with two rubber-insulated conductors in a flexible metal sheath were used as early as 1906, and were considered at the time a better method than open knob-and-tube wiring, although much more expensive.

The first polymer-insulated cables for building wiring were introduced in 1922. These were two or more solid copper electrical wires with rubber insulation, plus woven cotton cloth over each conductor for protection of the insulation, with an overall woven jacket, usually impregnated with tar as a protection from moisture. Waxed paper was used as a filler and separator.

Over time, rubber-insulated cables become brittle because of exposure to atmospheric oxygen, so they must be handled with care, and are usually replaced during renovations. When switches, outlets or light fixtures are replaced, the mere act of tightening connections may cause hardened insulation to flake off the conductors. Rubber insulation further inside the cable often is in better condition than the insulation exposed at connections, due to reduced exposure to oxygen.

Rubber insulation was hard to strip from bare copper, so copper was tinned, causing slightly more electrical resistance. Rubber insulation is no longer used for permanent wiring installations, but may still be used for replaceable temporary cables where flexibility is important, such as electrical extension cords.

About 1950, PVC insulation and jackets were introduced, especially for residential wiring. About the same time, single conductors with a thinner PVC insulation and a thin nylon jacket (e.g. US Type THN, THHN, etc.) became common.

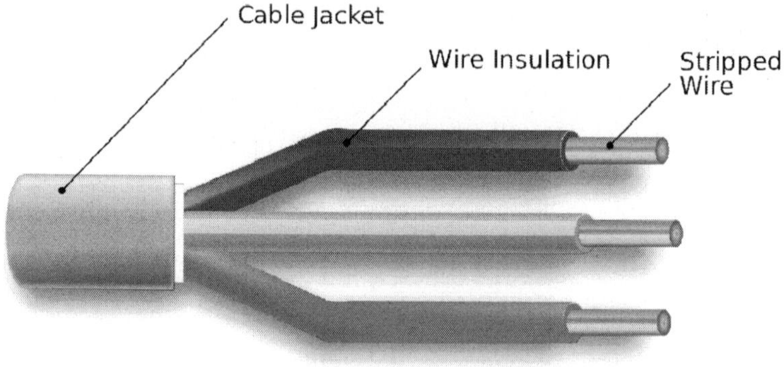

Figure: *Diagram of a simple electrical cable with three insulated conductors*

The simplest form of cable has two insulated conductors twisted together to form a unit; such unjacketed cables with two or three conductors are used for low-voltage signal and control applications such as doorbell wiring. In North American practice, an overhead cable from a transformer on a power pole to a residential electrical service consists of three twisted (triplexed) wires, often with one being a bare wire made of copper (protective earth/ground) and the other two being insulated for the line voltage (hot/live wire and neutral wire). For additional safety, the ground wire may be formed into a stranded co-axial layer completely surrounding the phase conductors, so that the outmost conductor is grounded.

Copper Conductors

Electrical devices often contain copper conductors because of their multiple beneficial properties, including their high electrical conductivity, tensile strength, ductility, creep resistance, corrosion resistance, thermal conductivity, coefficient of thermal expansion, solderability, resistance to electrical overloads, compatibility with electrical insulators, and ease of installation.

Despite competition from other materials, copper remains the preferred electrical conductor in nearly all categories of electrical wiring. For example, copper is used to conduct electricity in high, medium and low voltage power networks, including power generation, power transmission, power distribution, telecommunications, electronics circuitry, data processing, instrumentation, appliances, entertainment systems, motors, transformers, heavy industrial machinery, and countless other types of electrical equipment.

Aluminium Conductors

Aluminium wire was common in North American residential wiring from the late 1960s to mid-1970s due to the rising cost of copper. Because of its greater resistivity, aluminium wiring requires larger conductors than copper. For instance, instead of 14 AWG (American wire gauge) for most lighting circuits, aluminium wiring would be 12 AWG on a typical 15 ampere circuit, though local building codes may vary. Aluminium conductors were originally indiscriminately used with wiring devices intended for copper conductors. This practice was found to cause defective connections unless the aluminium was one of a special alloy, or all devices — breakers, switches, receptacles, splice connectors, wire nuts, etc. — were specially designed for the purpose. These special designs address problems with junctions between dissimilar metals, oxidation on metal surfaces, and mechanical effects that occur as different metals expand at different rates with increases in temperature.

Unlike copper, aluminium has a tendency to cold-flow under pressure, so screw clamped connections may get loose over time. This can be mitigated by using spring-loaded connectors that apply constant pressure, applying high pressure cold joints in splices and termination fittings, and torquing the bolted connection.

Also unlike copper, aluminium forms an insulating oxide layer on the surface. This is sometimes addressed by coating aluminium wires with an antioxidant paste at joints, or by applying a mechanical termination designed to break through the oxide layer during installation. Because of improper design and installation, some junctions to wiring devices would overheat under heavy current load, and cause fires. Revised standards for wiring devices (such as the CO/ALR "copper-aluminium-revised" designation) were developed to reduce these problems. Nonetheless, aluminium wiring for residential use has acquired a poor reputation and has fallen out of favour.

Aluminium conductors are still used for bulk power distribution and large feeder circuits, because they cost less than copper wiring, and weigh less, especially in the large sizes needed for heavy current loads. Aluminium conductors must be installed with compatible connectors.

Modern Wiring Materials

Modern non-metallic sheathed cables, such as (US and Canadian) Type NMB and NMC, consist of two to four wires covered with thermoplastic insulation and a bare wire for grounding (bonding) surrounded by a flexible plastic jacket. Some versions wrap the individual conductors in paper before the plastic jacket is applied. Romex brand NM cable is the most well-known brand of NM cable, since the first of its type was manufactured by Rome Cable Division of Cyprus Mines, Rome, New York. The Romex trademark has been owned by Southwire Company since it purchased the electrical building wire assets of General Cable in 2001.

Special versions of non-metallic sheathed cables, such as US Type UF, are designed for direct underground burial (often with separate mechanical protection) or exterior use where exposure to ultraviolet radiation (UV) is a possibility. These cables differ in having a moisture-resistant construction, lacking paper or other absorbent fillers, and being formulated for UV resistance.

Rubber-like synthetic polymer insulation is used in industrial cables and power cables installed underground because of its superior moisture resistance. Insulated cables are rated by their allowable operating voltage and their maximum operating temperature at the conductor surface.

A cable may carry multiple usage ratings for applications, for example, one rating for dry installations and another when exposed to moisture or oil.

Generally, single conductor building wire in small sizes is solid wire, since the wiring is not required to be very flexible. Building wire conductors larger than 10 AWG (or about 6 mm^2) are stranded for flexibility during installation, but not stranded enough to be flexible enough to use as appliance cord.

Cables for industrial, commercial, and apartment buildings may contain many insulated conductors in an overall jacket, with helical tape steel or aluminium armour, or steel wire armour, and perhaps as well an overall PVC or lead jacket for protection from moisture and physical damage. Cables intended for very flexible service or in marine applications may be protected by woven bronze wires. Power or communications cables (e.g., computer networking) that are routed in or through air-handling spaces (plenums) of office buildings are required under the model code to be either encased in metal conduit or rated for low flame and smoke production.

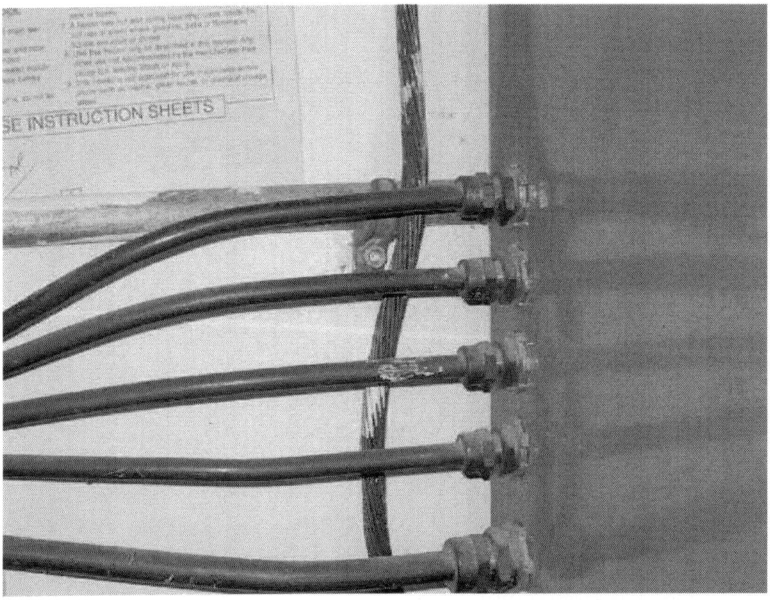

Figure: *Mineral insulated cables at a panel board*

For some industrial uses in steel mills and similar hot environments, no organic material gives satisfactory service. Cables insulated with compressed mica flakes are sometimes used. Another form of high-temperature cable is a mineral insulated cable, with individual conductors placed within a copper tube, and the space filled with magnesium oxide powder. The whole assembly is drawn down to smaller sizes, thereby compressing the powder. Such cables have a certified fire resistance rating, are more costly than non-fire rated cable, and have little flexibility and are effectively rigid to the user of the cable.

Because multiple conductors bundled in a cable cannot dissipate heat as easily as single insulated conductors, those circuits are always rated at a lower "ampacity". Tables in electrical safety codes give the maximum allowable current for a particular size of conductor, for the voltage and temperature rating at the surface of the conductor for a given physical environment, including the insulation type and thickness. The allowable current will be different for wet or dry, for hot (attic) or cool (underground) locations. In a run of cable through several areas, the most severe area will determine the appropriate rating of the overall run. Cables usually are secured by special fittings where they enter electrical apparatus; this may be a simple screw clamp for jacketed cables in a dry location, or a polymer-gasketed cable connector that mechanically engages the armour of an armoured cable and provides a water-resistant connection. Special cable fittings may be applied to prevent explosive gases from flowing in the interior of jacketed cables, where the cable passes through areas where inflammable gases are present. To prevent loosening of the connections of individual conductors of a cable, cables must be supported near their entrance to devices and at regular intervals through their length. In tall buildings special designs are required to support the conductors of vertical runs of cable. Usually, only one cable per fitting is allowed unless the fitting is otherwise rated.

Special cable constructions and termination techniques are required for cables installed in ocean-going vessels; in addition to electrical safety and fire safety, such cables may also be required to be pressure-resistant where they penetrate bulkheads of a ship.

Raceways

Insulated wires may be run in one of several forms of a raceway between electrical devices. This may be a specialised bendable pipe, called a conduit, or one of several varieties of metal (rigid steel or aluminium) or non-metallic (PVC or HDPE) tubing. Rectangular cross-section metal or PVC wire troughs (North America) or trunking (UK) may be used if many circuits are required. Wires run underground may be run in plastic tubing encased in concrete, but metal elbows may be used in severe pulls. Wiring in exposed areas, for example factory floors, may be run in cable trays or rectangular raceways having lids.

Where wiring, or raceways that hold the wiring, must traverse fire-resistance rated walls and floors, the openings are required by local building codes to be firestopped. In cases where safety-critical wiring must be kept operational during an accidental fire, fireproofing must be applied to maintain circuit integrity in a manner to comply with a product's certification listing. The nature and thickness of any passive fire protection materials used in conjunction with wiring and raceways has a quantifiable impact upon the ampacityderating, because the thermal insulation properties needed for fire resistance also inhibit air cooling of power conductors.

Cable trays are used in industrial areas where many insulated cables are run together. Individual cables can exit the tray at any point, simplifying the wiring installation and reducing the labour cost for installing new cables. Power cables may have fittings in the tray to maintain clearance between the conductors, but small control wiring is often installed without any intentional spacing between cables.

Since wires run in conduits or underground cannot dissipate heat as easily as in open air, and since adjacent circuits contribute induced currents, wiring regulations give rules to establish the current capacity (ampacity). Special sealed fittings are used for wiring routed through potentially explosive atmospheres.

Bus Bars, Bus Duct, Cable Bus

For very heavy currents in electrical apparatus, and for heavy currents distributed through a building, bus bars can be used. (The term "bus" is a contraction of the Latin *omnibus* - meaning "for all".) Each live conductor of such a system is a rigid piece of copper or aluminium, usually in flat bars (but sometimes as tubing or other shapes). Open bus bars are never used in publicly accessible areas, although they are used in manufacturing plants and power company switch yards to gain the benefit of air cooling. A variation is to use heavy cables, especially where it is desirable to transpose or "roll" phases.

In industrial applications, conductor bars are often pre-assembled with insulators in grounded enclosures. This assembly, known as bus duct or busway, can be used for connections to large switchgear or for bringing the main power feed into a building. A form of bus duct known as "plug-in bus" is used to distribute power down the length of a building; it is constructed to allow tap-off switches or motor controllers to be installed at designated places along the bus. The big advantage of this scheme is the ability to remove or add a branch circuit without removing voltage from the whole duct.

Bus ducts may have all phase conductors in the same enclosure (non-isolated bus), or may have each conductor separated by a grounded barrier from the adjacent phases (segregated bus). For conducting large currents between devices, a cable bus is used.

For very large currents in generating stations or substations, where it is difficult to provide circuit protection, an isolated-phase bus is used. Each phase of the circuit is run in a separate grounded metal enclosure. The only fault possible is a phase-to-ground fault, since the enclosures are separated. This type of bus can be rated up to 50,000 amperes and up to hundreds of kilovolts (during normal service, not just for faults), but is not used for building wiring in the conventional sense.

4

Building Insulation

Building insulation refers broadly to any object in a building used as insulation for any purpose. While the majority of insulation in buildings is for thermal purposes, the term also applies to acoustic insulation, fire insulation, and impact insulation (e.g. for vibrations caused by industrial applications). Often an insulation material will be chosen for its ability to perform several of these functions at once.

Thermal Insulation

Thermal insulation in buildings is an important factor to achieving thermal comfort for its occupants. Insulation reduces unwanted heat loss or gain and can decrease the energy demands of heating and cooling systems. It does not necessarily deal with issues of adequate ventilation and may or may not affect the level of sound insulation. In a narrow sense insulation can just refer to the insulation materials employed to slow heat loss, such as: cellulose, glass wool, rock wool, polystyrene, urethane foam, vermiculite, perlite, wood fibre, plant fibre (cannabis, flax, cotton, cork, etc.), plant straw, animal fibre (sheep's wool), cement, and earth or soil, but it can also involve a range of designs and techniques to address the main modes of heat transfer - conduction, radiation and convection materials.

The effectiveness of insulation is commonly evaluated by its R-value, of which there are two - metric (SI) and US customary, the former being 0.176 times the latter. For attics, it is recommended that it should be at least R-38 (US customary, R-6.7 metric). However, an R-value does not take into account the quality of construction or local environmental factors for each building. Construction quality issues include inadequate vapor barriers, and problems with draft-proofing.

In addition, the properties and density of the insulation material itself is critical.

Planning

How much insulation a house should have depends on building design, climate, energy costs, budget, and personal preference. Regional climates make for different requirements. Building codes specify only the bare minimum; insulating beyond what the code requires is often recommended.

The insulation strategy of a building needs to be based on a careful consideration of the mode of energy transfer and the direction and intensity in which it moves. This may alter throughout the day and from season to season. It is important to choose an appropriate design, the correct combination of materials and building techniques to suit the particular situation. To determine whether you should add insulation, you first need to find out how much insulation you already have in your home and where. A qualified home energy auditor will include an insulation check as a routine part of a whole-house energy audit.

In the USA

An initial estimate of insulation needs in the United States can be determined by the US Department of Energy's ZIP codeinsulation calculator.

Russia

In Russia the luxury of cheap gas has led to poorly insulated, overheated and inefficient consumers of energy. The Russian Centre for Energy Efficiency found that Russian buildings are either over- or under-heated, often consume up to 50 percent more heat and hot water than needed. Fifty-three percent of all carbon dioxide (CO_2) emissions in Russia are produced through heating and generating electricity for buildings. However, Green House Gas (GHG) emissions from the Soviet Bloc are still below their 1990 levels).

Climate

Cold climates: In cold conditions, the main aim is to reduce heat flow out of the building. The components of the building envelope - windows, doors, roofs, walls, and air infiltration barriers - are all important sources of heat loss; in an otherwise well insulated home, windows will then become an important source of heat transfer. The resistance to conducted heat loss for standard glazing corresponds to an R-value of about 0.17W/m²/K° (compared to 2-4W/m²/K° for glasswoolbatts). Losses can be reduced by good weatherisation, bulk insulation, and minimising the amount of non-insulative (particularly non-solar facing) glazing. Indoor thermal radiation can also be a disadvantage with spectrally selective (low-e, low-emissivity) glazing. Some insulated glazing systems can double to triple R values.

Figure: *Cross-section of home insulation.*

Hot Climates

In hot conditions, the greatest source of heat energy is solar radiation. This can enter buildings directly through windows or it can heat the building shell to a higher temperature than the ambient, increasing the heat transfer through the building envelope. The Solar Heat Gain Co-efficient (SGHC) (a measure of solar heat transmittance) of standard single glazing can be around 78-85%. Solar gain can be reduced by adequate shading from the sun, light coloured roofing, spectrally selective (heat-reflective) paints and coatings and various types of insulation for the rest of the envelope. Specially coated glazing can reduce SHGC to around 10%. Radiant barriers are highly effective for attic spaces in hot climates. In this application, they are much more effective in hot climates than cold climates. For downward heat flow, convection is weak and radiation dominates heat transfer across an air space. Radiant barriers must face an adequate air-gap to be effective.

If refrigerative air-conditioning is employed in a hot, humid climate, then it is particularly important to seal the building envelope. Dehumidification of humid air infiltration can waste significant energy. On the other hand, some building designs are based on effective cross-ventilation instead of refrigerative air-conditioning to provide convective cooling from prevailing breezes.

Orientation - Passive Solar Design

Optimal placement of building elements (e.g. windows, doors, heaters) can play a significant role in insulation by considering the impact of solar radiation on the building and the prevailing breezes. Reflective laminates can help reduce passive solar heat in pole barns, garages and metal buildings.

Building Envelope

The thermal envelope defines the conditioned or living space in a house. The attic or basement may or may not be included in this area. Reducing airflow from inside to outside can help to reduce convective heat transfer significantly. Ensuring low convective heat transfer also requires attention to building construction (weatherization) and the correct installation of insulative materials.

The less natural airflow into a building, the more mechanical ventilation will be required to support human comfort. High humidity can be a significant issue associated with lack of airflow, causing condensation, rotting construction materials, and encouraging microbial growth such as mould and bacteria. Moisture can also drastically reduce the effectiveness of insulation by creating a thermal bridge. Air exchange systems can be actively or passively incorporated to address these problems.

Some of the unwanted continuous air exchange with the outside air, which is an artifact of the inherent hysterisis of thermostat control systems, is inevitable. These thermostats typically require the indoor air temperature to cycle 1 to 2 degrees Fahrenheit (about 1 Celsius) as the heating system cycles on and off. Each of these cycles results in a 0.2% to 0.4% volume exchange with the outside air as conditioned envelope air expands and contracts.

Thus a 3000 square foot home with 9 foot ceilings having a volume of 27,000 cubic feet and a heating/cooling system cycling every ten minutes will exchange 7700 to 15400 cubic feet in a 24 hr period. This incoming, unconditioned air places an additional load on heating and cooling systems and results in additional costs to energy consumers. Interestingly, the cycling frequency of the heating/cooling system is at maximum when the heating/cooling load equals half the heating/cooling system capacity.

Thermal Bridge

Thermal bridges are points in the building envelope that allow heat conduction to occur. Since heat flows through the path of least resistance, thermal bridges can contribute to poor energy performance. A thermal bridge is created when materials create a continuous path across a temperature difference, in which the heat flow is not interrupted by thermal insulation. Common building materials that are poor insulators include glass and metal.

A building design may have limited capacity for insulation in some areas of the structure. A common construction design is based on stud walls, in which thermal bridges are common in wood or steel studs and joists, which are typically fastened with metal. Notable areas that most commonly lack sufficient insulation are the corners of buildings, and areas where insulation has been removed or displaced to make room for system infrastructure, such as electrical boxes (outlets and light switches), plumbing, fire alarm equipment, etc.

Thermal bridges can also be created by uncoordinated construction, for example by closing off parts of external walls before they are fully insulated. The existence of inaccessible voids within the wall cavity which are devoid of insulation can be a source of thermal bridging.

Some forms of insulation transfer heat more readily when wet, and can therefore also form a thermal bridge in this state.

The heat conduction can be minimized by any of the following: reducing the cross sectional area of the bridges, increasing the bridge length, or decreasing the number of thermal bridges.

One method of reducing thermal bridge effects is the installation of an insulation board (e.g. foam board EPS XPS, wood fibre board, etc.) over the exterior outside wall. Another method is using insulated lumber framing for a thermal break inside the wall.

Materials

There are essentially two types of building insulation - Bulk Insulation and Reflective Insulation. Most buildings use a combination of both types to make up a total building insulation system. The type of insulation used is matched to create maximum resistance to each of the three forms of building heat transfer - Conduction, Convection, and Radiation.

Conductive and Convective Insulators ('Bulk Insulation')

Bulk insulators block conductive heat transfer and convective flow either into or out of a building. The denser a material is, the better it will conduct heat. Because air has such low density, air is a very poor conductor and therefore makes a good insulator. Insulation to resist conductive heat transfer uses air spaces between fibres, inside foam or plastic bubbles and in building cavities like the attic. This is beneficial in an actively cooled or heated building, but can be a liability in a passively cooled building; adequate provisions for cooling by ventilation or radiation are needed.

Radiant Heat Barriers

Radiant barriers work in conjunction with an air space to reduce radiant heat transfer across the air space. Radiant or reflective insulation reflects heat instead of either absorbing it or letting it pass through. Radiant barriers are often seen used in reducing downward heat flow, because upward heat flow tends to be dominated by convection. This means that for attics, ceilings, and roofs, they are most effective in hot climates. They also have a role in reducing heat losses in cool climates. However, much greater insulation can be achieved through the addition of bulk insulators.

Some radiant barriers are spectrally selective and will preferentially reduce the flow of infra-red radiation in comparison to other wavelengths. For instance low-emissivity (low-e) windows will transmit light and short-wave infra-red energy into a building but reflect back the long-wave infra-red radiation generated by interior furnishings. Similarly, special heat-reflective paints are able to reflect more heat than visible light, or vice-versa.

Thermal emissivity values probably best reflect the effectiveness of radiant barriers. Some manufacturers quote an 'equivalent' R-value for these products but these figures can be difficult to interpret, or even misleading, since R-value testing measures total heat loss in a laboratory setting and does not control the type of heat loss responsible for the net result (radiation, conduction, convection). A film of dirt or moisture can alter the emissivity and hence the performance of radiant barriers.

Installation of Insulation

Insulating buildings during construction is much easier than retrofitting, as generally the insulation is hidden, and parts of the building need to be deconstructed to reach them.

Wood Stain

A wood stain consists of a colourant suspended or dissolved in an agent or solvent. The suspension agent can be water, alcohol, petroleum distillate, or the actual finishing agent (shellac, lacquer, varnish, polyurethane, etc.). Coloured or 'stained' finishes, like polyurethane, do not penetrate the pores of the wood to any significant degree and will disappear when the finish itself deteriorates or is removed intentionally.

Pigments and dyes are largely used as colourants. The difference between the two is in the size of the particles. Dyes are microscopic crystals that dissolve in the vehicle and pigments are suspended in the vehicle and are much larger. Dyes will colour very fine grained wood, like cherry or maple, which pigments will not. Those fine-grained woods have pores too small for pigments to attach themselves to. Pigments contain a binder to help attach themselves to the wood.

The type of stain will either accentuate or obscure the wood grain. Most commercial stains contain both dye and pigment and the degree to which they stain the appropriate wood is mostly dependent on the length of time they are left on the wood. Pigments, regardless of the suspension agent, will not give much colour to very dense woods but will deeply colour woods with large pores (*e.g.* pine). Dyes are translucent and pigments are opaque.

Gel stains are more akin to paint and have little penetrating ability.

Composition

Stain is composed of the same three primary ingredients as paint (pigment, solvent, and binder) but is predominantly pigment (or dye) and solvent with little binder. Much like the dyeing or staining of fabric, wood stain is designed to add colour to the substrate of wood and other materials while leaving the substrate mostly visible. Transparent varnishes or surface films are applied afterwards. In principle, stains do not provide a surface coating or film. However, because the binders are from the same class of film-forming binders that are used in paints and varnishes, some build-up of film occurs.

Comparison with Paint and Varnish

The initial application of any paint or varnish is similarly absorbed into the substrate, but because stains contain lower amounts of binder, the binder from a stain resides mainly below the surface while the pigment remains near the top or at the surface. Stains that employ metallic pigments such as iron oxides usually are more opaque; first because metallic pigments are opaque by nature, but also because the particles of which they consist are much larger than organic pigments and therefore do not penetrate as well. Most wood stains for interior uses (e.g. floors and furniture) require further application of varnish or finish for protection and gloss. Stains are differentiated from varnishes in that the latter has no added colour or pigment and is designed to form a surface film. Some products are marketed as a combination of stain and varnish.

Siding Stain

Siding stain is one variety of wood stain with very high viscosity (others can be quite thin). Effectively, siding stains are paints that do not cover as well and do not form a hard film. They are designed to penetrate better and contain binders that are softer and more flexible, allowing them to last longer than harder, more brittle paints. Siding stain protects against solar radiation especially UV radiation, water, fungus including mildew, and insects. Different siding stains are distinguished by the appearance they impart to wood. Certain solvent-based or oil-based siding stains contain small amounts of paraffin wax, which cannot be painted over, although re-staining is still possible.

Absorption

Applying stains can be very easy or very difficult depending on the type of substrate, type of stain, and the ambient conditions. Fresh, "green" lumber accepts stain poorly, while aged wood absorbs stains relatively well. Porosity of wood can vary greatly, even within the same piece of wood. End grain and bias-cut grain are far more absorbent, thus will accept more pigment and will darken considerably in those areas. The hard ring may absorb differently from the soft ring. The characteristic medullary rays in oak will absorb much less and remain mostly blonde. Woods that have been heavily subjected to paint strippers or washed down with detergents or solvents will have an increased open grain and accept substantially more stain than normal. Woods from different species of trees can have huge variations in how well they take stain. Different wood species stain differently—the overall colour and shade is a result of a combination of the stain and properties of the wood. For example, although medium-to-dark stains tend to look blotchy on maple, they get deeper and more glowing on cherry, with a more consistent colouration. Stains that are fast drying will be difficult to apply in hot weather or in direct sunlight. Stains that are slow-drying will be difficult to work with in damp and cold conditions due to a greatly lengthened evaporation and curing period. New lumber, such as pine, can have waxlike sealants put on at the mill that will prevent proper staining; stripping or sanding the surface may be required. White stains composed of metal oxides, namely titanium dioxide and zinc oxide do not penetrate well and remain on the surface. In such cases, wear easily reveals unstained wood. They are also fairly opaque.

Preparation

Typically, a thorough final sanding of the wood with one suitable grit will "even out" the absorption of the stain. White stain on a bare softwood or oak floor might require a final 'prep' sanding by hand with an orbital/vibrating sander with 80 or 100 grit, whereas certain hardwoods might be orbitally or hand sanded with 220 grit and higher for a darkish organic stain on furniture. Though it is not as durable, "garnet" is by far the preferred sandpaper for hand-sanding bare wood; first because it is a sharper and faster cutting grit, and second because it does not impart colour from the grit, as aluminium oxide or silicon carbide can do. Other methods include "rubbing" with rotten-stone etc.

In certain cases you need to clean the wood or remove existing stains prior to staining the wood with a commercial stain in order to avoid damaging the wood. This can be the case for both unfinished and finished wood.

Cleaning Stained Wood

There are special consideration when cleaning stained wood. One of the most common stains is water stains on stained wood. Techniques to remove the water stains have been documented that use a hot iron to remove the water stain. Special caution should be used when trying to remove stains from a stained wood to avoid damaging the original stain. This is especially important with antiques.

Glass Brick

Glass brick, also known as glass block, is an architectural element made from glass. Glass bricks provide visual obscuration while admitting light. The glass block was originally developed in the early 1900s to provide natural light in manufacturing plants. Glass bricks are produced for both wall and floor applications. Glass blocks for use in floors are normally manufactured as a single solid piece, or as a hollow glass block with thicker side walls than the standard wall blocks. These blocks are normally cast into a reinforced concrete gridwork or set into a metal frame, allowing multiple units to be combined to span over openings in basements and roofs. Glass wall blocks should not be used in flooring applications. Hollow glass wall blocks are manufactured as two separate hemispheres and, whilst the glass is still molten, the two hemispheres are pressed together and annealed. The resulting glass blocks will have a partial vacuum at the hollow centre.

Specialist glass blocks are produced for various applications including:

Bullet and vandal resistance: these blocks are generally solid glass or have very thick side walls similar to pavement blocks.

Fire resistance: fire resistance of varying degrees can be achieved by several methods. Standard production hollow wall block will offer little fire resistance; however, resistance is improved by utilising specially produced hollow blocks with thicker sidewalls, or the inclusion of a special layer of fire resisting material between the two hemispheres of the block during manufacture. Some manufacturers of glass blocks have developed a method of bonding two glass blocks together with adhesive, producing blocks of up to 160 mm (6½") thick with enhanced fire resistance. It is important that the block manufacturer's recommendations are followed with regards to the installation of fire resisting glass block walls, as without special construction techniques, the wall will not achieve the desired fire resistance.

A recent innovation in the manufacture of glass blocks is the inclusion of argon gas within the hollow centre of glass wall blocks. This advancement in production technique has resulted in a glass block which is able to offer significantly improved thermal insulation properties.

Glass blocks in Europe are manufactured in accordance with the European Standard EN1052-2. The International Standard is ISO TC 160/SG1. The Standards allow for variation in sizes and production irregularity. Blocks fall within three classifications; Class1, Class 2 and Class 3 with Class 1 being the highest and best rating with a maximum permissible deviation from designed size and rectangularity of 1 mm. Some hollow glass wall blocks are available in coloured variants. These coloured variants fall into two categories; those that are manufactured with coloured glass which are UV stable and can be used in the same locations as standard clear glass blocks.

The other method by which coloured glass blocks are achieved is to inject a coloured material, dye or transparent paint into the hollow centre of the blocks to form a permanent coating. This method of producing coloured blocks enables vibrant colours to be achieved which are not possible with coloured glass. The downside of this production method is that the coloured coating may not be UV stable and can fade in bright sunshine over time and may therefore, not be suitable for all locations.

Glass wall blocks are fixed together to form complete walls by several methods – the most common method of construction is to bed the blocks together in a Portland cement-based mortar with reinforcing rods of steel placed within the mortar as recommended by the project architect or block manufacturer. Other methods of construction include several proprietary systems whereby the mortar is replaced by timber or PVC extrusions.

Structural Steel

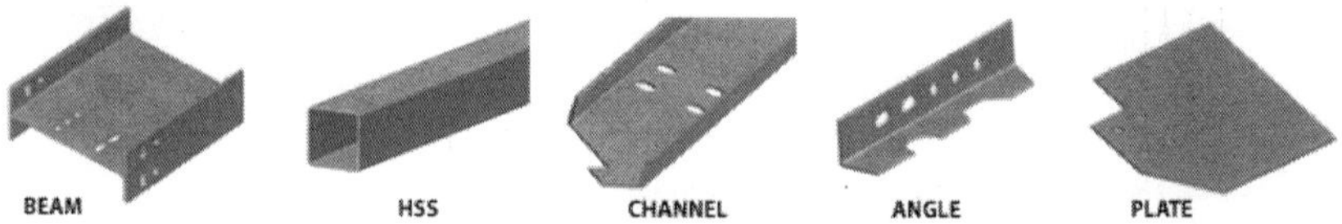

Figure: *Various structural steel shapes*

Structural steel is steel construction material, a profile, formed with a specific shape or cross section and certain standards of chemical composition and mechanical properties. Structural steel shape, size, composition, strength, storage, etc., is regulated in most industrialized countries. Structural steel members, such as I-beams, have high second moments of area, which allow them to be very stiff in respect to their cross-sectional area.

Common Structural Shapes

In most developed countries, the shapes available are set out in published standards, although a number of specialist and proprietary cross sections are also available.

- I-beam (I-shaped cross-section - in Britain these include Universal Beams (UB) and Universal Columns (UC); in Europe it includes the IPE, HE, HL, HD and other sections; in the US it includes Wide Flange (WF) and H sections)
- Z-Shape (half a flange in opposite directions)
- HSS-Shape (Hollow structural section also known as SHS (structural hollow section) and including square, rectangular, circular (pipe) and elliptical cross sections)
- Angle (L-shaped cross-section)
- Channel, or C section, ([-shaped cross-section)
- Tee (T-shaped cross-section)
- Rail profile (asymmetrical I-beam)
  - Railway rail
  - Vignoles rail
  - Flanged T rail
  - Grooved rail
- Bar, a piece of metal, rectangular cross sectioned (flat) and long, but not so wide so as to be called a sheet.
- Rod, a round or square and long piece of metal or wood.
- Plate, metal sheets thicker than 6 mm or $^1/_4$ in.
- Open web steel joist

While many sections are made by hot or coldrolling, others are made by welding together flat or bent plates (for example, the largest circular hollow sections are made from flat plate bent into a circle and seam-welded).

Standards

Standard Structural Steels (Europe): Most steels used throughout Europe are specified to comply with the European standard EN 10025. However, many national standards also remain in force.

Typical grades are described as 'S275J2' or 'S355K2W'. In these examples, 'S' denotes structural rather than engineering steel; 275 or 355 denotes the yield strength in newtons per square millimetre or the equivalent megapascals; J2 or K2 denotes the materials toughness by reference to Charpy impact test values; and the 'W' denotes weathering steel. Further letters can be used to designate fine grain steel ('N' or

'NL'); quenched and tempered steel ('Q' or 'QL'); and thermomechanically rolled steel ('M' or 'ML'). The normal yield strength grades available are 195, 235, 275, 355, 420, and 460, although some grades are more commonly used than others e.g. in the UK, almost all structural steel is grades S275 and S355. Higher grades are available in quenched and tempered material (500, 550, 620, 690, 890 and 960 - although grades above 690 receive little if any use in construction at present).

A set of euronorms define the shape of a set of standard structural profiles:

- European I-beam: IPE - Euronorm 19-57
- European I-beam: IPN - DIN 1025-1
- European flange beams: HE - Euronorm 53-62
- European channels: UPN - DIN 1026-1
- European cold formed IS IS 800-1

Standard Structural Steels (USA)

Steels used for building construction in the US use standard alloys identified and specified by ASTM International. These steels have an alloy identification beginning with A and then two, three, or four numbers. The four-number AISI steel grades commonly used for mechanical engineering, machines, and vehicles are a completely different specification series. The standard commonly used structural steels are:

Carbon Steels:

- A36 - structural shapes and plate
- A53 - structural pipe and tubing
- A500 - structural pipe and tubing
- A501 - structural pipe and tubing
- A529 - structural shapes and plate

High Strength Low Alloy Steels:

- A441 - structural shapes and plates
- A572 - structural shapes and plates
- A618 - structural pipe and tubing
- A992 - Possible applications are W or S I-Beams.
- A270 - structural shapes and plates

Corrosion Resistant High Strength Low Alloy Steels:

- A242 - structural shapes and plates
- A588 - structural shapes and plates

Quenched and Tempered Alloy Steels:

- A514 - structural shapes and plates
- A517 - boilers and pressure vessels

CE Marking

The concept of CE marking for all construction products and steel products is introduced by the Construction Products Directive (CPD). The CPD is a European Directive that ensures the free movement of all construction products within the European Union.

Because steel components are "safety critical", CE Marking is not allowed unless the Factory Production Control (FPC) system under which they are produced has been assessed by a suitable certification body that has been approved to the European Commission.

In the case of steel products such as sections, bolts and fabricated steelwork the CE Marking demonstrates that the product complies with the relevant harmonized standard.

For steel structures the main harmonized standards are:

- Steel sections and plate - EN 10025-1
- Hollow sections - EN 10219-1 and EN 10210-1
- Pre-loadable bolts - EN 14399-1
- Non-preloadable bolts - EN 15048-1
- Fabricated steel - EN EN 1090 -1

The standard that covers CE Marking of structural steelwork is EN 1090-1. The standard has come into force in late 2010. After a transition period of two years, CE Marking will become mandatory in most European Countries sometime early in 2012. The official end date of the transition period is July 1, 2014.!

Steel vs. Concrete

Choosing the Ideal Structural Material: Most construction projects require the use of hundreds of different materials. These range from concrete of all different specifications, structural steel of different specifications, clay, mortar, ceramics, wood, etc. In terms of a load bearing structural frame, they will generally consist of structural steel, concrete, masonry, and/or wood, using an infinite combination of each to produce an efficient structure. Most commercial and industrial structures are primarily constructed using either structural steel or reinforced concrete. When designing a structure, an engineer must decide which, if not both, material is most suitable for the design. There are many factors considered when choosing a construction material. Cost is commonly the controlling element; however, other considerations such as weight, strength, constructability, availability, sustainability, and fire resistance will be taken into account before a final decision is made.

- Cost - The cost of these construction materials will depend entirely on the geographical location of the project and the availability of the materials. Just as price gasoline fluctuates, so does cement, aggregate, steel, etc. Reinforced concrete attributes about half of its construction costs to the required form-work. This refers to the lumber necessary to build the "box" or container to pour the concrete into place and remain there until it cures. Due to this high cost, precast concrete is becoming a popular option for designers so costs can be saved, as well as time. Structural steel costs will depend most entirely on the amount of steel used since steel is sold by the pound. It is therefore the job of the structural designer to specify the lightest steel members possible while still maintaining a safe structural design. Another method of reducing costs in steel design is to use many of the same size steel members, as opposed to many unique members.
- Strength/weight ratio - Construction materials are commonly categorized by their strength to weight ratio, or specific strength. This is defined as the strength of a material over its density. This gives an engineer an indication as to how useful the material is in comparison to its weight, with the weight being a direct indication of its cost (typically) and ease of construction. Concrete is typically ten times stronger in compression than in tension, giving it a higher strength to weight ratio in compression, only.
- Sustainability - Many construction companies and material vendors are making changes to be a more environmentally friendly company. Sustainability has become an entirely new consideration for materials that are to be placed into the environment for generations of time. A sustainable material will be one that has minimal effect on the environment, both at the time of installation as well as throughout the life cycle of the material. Reinforced concrete and structural steel both have the ability to be a sustainable construction option, if used properly. Over 80% of structural steel members fabricated today come from recycled metals, called A992 steel. This

member material is cheaper, as well as having a higher strength to weight ratio, than previously used steel members (A36 grade). Reinforced concrete can also be a very sustainable option. Concrete's material components are naturally occurring materials that are not harmful to the environment. Also, concrete can now be poured to be permeable, allowing the flow of water through a paved surface and therefore reducing the necessary infrastructural construction that causes further potential disturbance to the environment. Concrete can also be crushed and used as aggregate in future concrete applications, meaning it doesn't necessarily have to be used as a land fill.

- Fire resistance - One of the most dangerous hazards to a building is a fire hazard. This is especially true in dry, windy climates and for structures constructed using wood. Special considerations must be taken into account with structural steel to ensure it is not under a dangerous fire hazard condition. Reinforced concrete characteristically does not pose a threat in the event of fire and even resists the spreading of fire, as well as temperature changes. This makes concrete an excellent insulation, improving the sustainability of the building it surrounds by reducing the required energy to maintain climate.
- Corrosion - When choosing a structural material, it is important to consider the life cycle of the building. Some materials are susceptible to corrosion from their surrounding elements, such as water, heat, humidity, or salt. Special considerations must be taken into account during the installation of a structural material to prevent any potential corrosion hazards. This must also be made clear to the occupants of the building because there may or may not be a necessary maintenance requirement to prevent corrosion. For example, structural steel cannot be exposed to the environment because any moisture, or other contact with water, will cause it to rust. When the steel rusts it compromises the structural integrity of the building and poses a potential danger to the residual or surrounding occupants.

Reinforced Concrete

- Characteristics - Generally consisting of portland cement, water, construction aggregate (coarse and fine), and steel reinforcing bars (rebar), concrete is cheaper in comparison to structural steel.

- Strength - Concrete is a composite material with relatively high compressive strength properties, but lacking in tensile strength/ductility. This inherently makes concrete a useful material for carrying the weight of a structure. Concrete reinforced with steel rebar give the structure a stronger tensile capacity, as well as an increase in ductility and elasticity.
- Constructability - Reinforced concrete must be poured and left to set, or harden. After setting (typically 1–2 days), a concrete must cure, the process in which concrete experiences a chemical reaction between the cementitious particles and the water. The curing process is complete after 28 days; however, construction may continue after 1–2 weeks, depending on the nature of the structure. Concrete can be constructed into nearly any shape and size. Approximately half of the cost of using reinforced concrete in a structural project is attributed to the construction of the form-work. In order to save time, and therefore costs, structural concrete members may be pre-cast. This refers to a reinforced concrete beam, girder, or column being poured off site and left to cure. After the curing process, the concrete member may be delivered to the construction site and installed as soon as it is needed. Since the concrete member was cured off location beforehand, construction may continue immediately after erection.
- Fire resistance - Concrete has excellent fire resistance properties, requiring no additional construction costs to adhere to the International Building Code (IBC) fire protection standards. However, concrete buildings will still likely use other materials that are not fire resistant. Therefore, a designer must still take into account the use of the concrete and where it will require fire hazardous materials in order to prevent future complications in the overall design.
- Corrosion - Reinforced concrete, when constructed properly, has excellent corrosion resistance properties. Concrete is not only resistant to water, but needs it to cure and develop its strength over time. However, the steel reinforcement in the concrete must not be exposed in order to prevent its corrosion as this could significantly reduce the ultimate strength of the structure. The American Concrete Institute provides the necessary design specifications for an engineer to ensure there is enough concrete covering any steel reinforcement to prevent exposure to water. This cover distance must be specified because

concrete will inevitable crack at locations carrying tension, or locations containing reinforcing bars for the purpose of carrying said tension. If the concrete cracks, it provides a path for water to travel directly to the reinforcing bars. Some reinforcing bars are coated in epoxy as a second order measure of preventing corrosion due to water contact. This method induces higher costs on the overall project, however, due to the higher cost of the epoxy coated bars. Also, when using epoxy coated bars, reinforced concrete members must be designed larger, as well as stronger, in order to balance the loss of friction between the reinforcing bars and concrete. This friction is referred to as bond strength, and it is vital to the structural integrity of a concrete member.

Structural Steel

- Characteristics - Structural steel differs from concrete in its attributed compressive strength as well as tensile strength.
- Strength - Having high strength, stiffness, toughness, and ductile properties, structural steel is one of the most commonly used materials in commercial and industrial building construction.
- Constructability - Structural steel can be developed into nearly any shape, which are either bolted or welded together in construction. Structural steel can be erected as soon as the materials are delivered on site, whereas concrete must be cured at least 1–2 weeks after pouring before construction can continue, making steel a schedule-friendly construction material.
- Fire resistance - Steel is inherently a noncombustible material. However, when heated to temperatures seen in a fire scenario, the strength and stiffness of the material is significantly reduced. The International Building Code requires steel be enveloped in sufficient fire-resistant materials, increasing overall cost of steel structure buildings.
- Corrosion - Steel, when in contact with water, can corrode, creating a potentially dangerous structure. Measures must be taken in structural steel construction to prevent any lifetime corrosion. The steel can be painted, providing water resistance. Also, the fire resistance material used to envelope steel is commonly water resistant.

The tallest structures today (commonly called "skyscrapers" or high-rise) are constructed using structural steel due to its constructability, as well as its high strength-to-weight ratio. In comparison, concrete, while being less dense than steel, has a much lower strength-to-weight ratio. This is due to the much larger volume required for a structural concrete member to support the same load; steel, though denser, does not require as much material to carry a load. However, this advantage becomes insignificant for low-rise buildings, or those with several stories or less. Low-rise buildings distribute much smaller loads than high-rise structures, making concrete the economical choice. This is especially true for simple structures, such as parking garages, or any building that is a simple, rectilinear shape.

Structural steel and reinforced concrete are not always chosen solely because they are the most ideal material for the structure. Companies rely on the ability to turn a profit for any construction project, as do the designers. The price of raw materials (steel, cement, coarse aggregate, fine aggregate, lumber (for form-work), etc., is constantly changing. If a structure could be constructed using either material, the cheapest of the two will likely control. Another significant variable is the location of the project. The closest steel fabrication facility may be much further from the construction site than the nearest concrete supplier. The high cost of energy and transportation will control the selection of the material as well. All of these costs will be taken into consideration before the conceptual design of a construction project is begun.

Combining Steel and Reinforced Concrete

Structures consisting of both materials utilise the benefits of structural steel and reinforced concrete. This is already common practice in reinforced concrete in that the steel reinforcement is used to provide steel's tensile strength capacity to a structural concrete member. A commonly seen example would be parking garages. Some parking garages are constructed using structural steel columns and reinforced concrete slabs. The concrete will be poured for the foundational footings, giving the parking garage a surface to be built on. The steel columns will be connected to the slab by bolting and/ or welding them to steel studs extruding from the surface of the poured concrete slab. Pre-cast concrete beams may be delivered on site to be installed for the second floor, after which a concrete slab may be poured for the pavement area. This can be done for multiple stories. A parking garage of this type is just one possible example of many structures that may use both reinforced concrete and structural steel.

A structural engineer understands that there are an infinite number of designs that will produce an efficient, safe, and affordable building. It is the engineer's job to work alongside the owner(s), contractor(s), and all other parties involved to produce an ideal product that suits everyone's needs. When choosing the structural materials for their structure, the engineer has many variables to consider, such as the cost, strength/weight ratio, sustainability of the material, constructability, etc.

Thermal Properties

The properties of steel vary widely, depending on its alloying elements. The austenizing temperature, the temperature where a steel transforms to an austenite crystal structure, for steel starts at 900 °C (1,650 °F) for pure iron, then, as more carbon is added, the temperature falls to a minimum 724 °C (1,335 °F) for eutectic steel (steel with only .83% by weight of carbon in it). As 2.1% carbon (by mass) is approached, the austenizing temperature climbs back up, to 1,130 °C (2,070 °F). Similarly, the melting point of steel changes based on the alloy.

The lowest temperature at which a plain carbon steel can begin to melt, its solidus, is 1,130 °C (2,070 °F). Steel never turns into a liquid below this temperature. Pure Iron ('Steel' with 0% Carbon) starts to melt at 1,492 °C (2,718 °F), and is completely liquid upon reaching 1,539 °C (2,802 °F). Steel with 2.1% Carbon by weight begins melting at 1,130 °C (2,070 °F), and is completely molten upon reaching 1,315 °C (2,399 °F). 'Steel' with more than 2.1% Carbon is no longer Steel, but is known as Cast iron.

Fire Resistance

Steel loses strength when heated sufficiently. The *critical temperature* of a steel member is the temperature at which it cannot safely support its load. Building codes and structural engineering standard practice defines different critical temperatures depending on the structural element type, configuration, orientation, and loading characteristics. The critical temperature is often considered the temperature at which its yield stress has been reduced to 60% of the room temperature yield stress. In order to determine the fire resistance rating of a steel member, accepted calculations practice can be used, or a fire test can be performed, the critical temperature of which is set by the standard accepted to the Authority Having Jurisdiction, such as a building code. In Japan, this is below 400°C. In China, Europe and North America (e.g., ASTM E-119), this is approximately

1000–1300F (530-810C). The time it takes for the steel element that is being tested to reach the temperature set by the test standard determines the duration of the fire-resistance rating. Heat transfer to the steel can be slowed by the use of fireproofing materials, thus limiting steel temperature. Common fireproofing methods for structural steel include intumescent, endothermic and plaster coatings as well as drywall, calcium silicate cladding, and mineral or high temperature insulation mineral wool blanket.

Concrete building structures often meet code required fire-resistance ratings, as the concrete thickness over the steel rebar provides sufficient fire resistance. However, concrete can be subject to spalling, particularly if it has an elevated moisture content. Although additional fireproofing is not often applied to concrete building structures, it is sometimes used in traffic tunnels and locations where a hydrocarbon fuel fire is more likely, as flammable liquid fires provides more heat to the structural element as compared to a fire involving ordinary combustibles during the same fire period. Structural steel fireproofing materials include intumescent, endothermic and plaster coatings as well as drywall, calcium silicate cladding, and mineral or high temperature insulation wool blankets. Attention is given to connections, as the thermal expansion of structural elements can compromise fire-resistance rated assemblies.

Manufacturing

Cutting workpieces to length is usually done with a bandsaw. A beam drill line (drill line) has long been considered an indispensable way to drill holes and mill slots into beams, channels and HSS elements. CNC beam drill lines are typically equipped with feed conveyors and position sensors to move the element into position for drilling, plus probing capability to determine the precise location where the hole or slot is to be cut.

For cutting irregular openings or non-uniform ends on dimensional (non-plate) elements, a cutting torch is typically used. Oxy-fuel torches are the most common technology and range from simple hand-held torches to automated CNC coping machines that move the torch head around the structural element in accordance with cutting instructions programmed into the machine.

Fabricating flat plate is performed on a plate processing centre where the plate is laid flat on a stationary 'table' and different cutting heads traverse the plate from a gantry-style arm or "bridge." The cutting heads can include a punch, drill or torch.

I-Beam

An I-beam, also known as H-beam, W-beam (for "wide flange"), Universal Beam (UB), Rolled Steel Joist (RSJ), or double-T (especially in Polish, Bulgarian, Spanish, Italian and German), is a beam with an I- or H-shaped cross-section. The horizontal elements of the "I" are flanges, while the vertical element is termed the "web".

The web resists shear forces, while the flanges resist most of the bending moment experienced by the beam. Beam theory shows that the I-shaped section is a very efficient form for carrying both bending and shear loads in the plane of the web. On the other hand, the cross-section has a reduced capacity in the transverse direction, and is also inefficient in carrying torsion, for which hollow structural sections are often preferred.

History

The method of producing an I-beam, as rolled from a single piece of steel, was patented by Alphonse Halbou of the company *Forges de la Providence* in 1849. Bethlehem Steel was a leading supplier of rolled structural steel of various cross-sections in American bridge and skyscraper work of the mid-twentieth century. Today, rolled cross-sections have been partially displaced in such work by fabricated cross-sections.

Overview

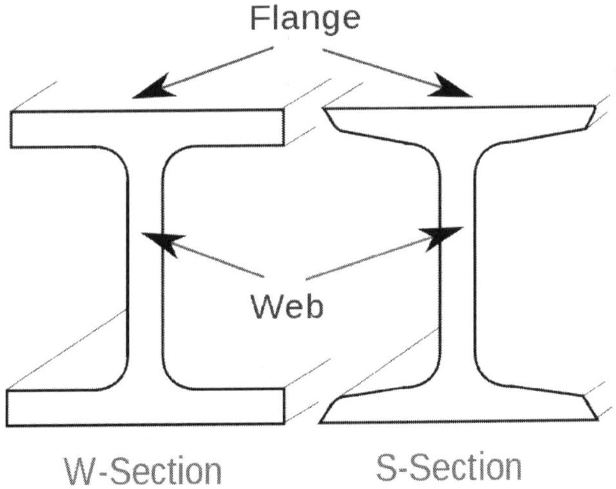

Figure: *Typical cross-sections of I-beams.*

There are two standard I-beam forms:

- Rolled I-beam, formed by hot rolling, cold rolling or extrusion (depending on material).

- Plate girder, formed by welding (or occasionally bolting or riveting) plates.

I-beams are commonly made of structural steel but may also be formed from aluminium or other materials. A common type of I-beam is the *rolled steel joist* (RSJ)—sometimes incorrectly rendered as *reinforced steel joist*. British and European standards also specify Universal Beams (UBs) and Universal Columns (UCs). These sections have parallel flanges, as opposed to the varying thickness of RSJ flanges which are seldom now rolled in the UK. Parallel flanges are easier to connect to and do away with the need for tapering washers. UCs have equal or near-equal width and depth and are more suited to being orientated vertically to carry axial load such as columns in multi-storey construction, while UBs are significantly deeper than they are wide are more suited to carrying bending load such as beam elements in floors.

I-joists — I-beams engineered from wood with fibreboard and/or laminated veneer lumber — are also becoming increasingly popular in construction, especially residential, as they are both lighter and less prone to warping than solid wooden joists. However there has been some concern as to their rapid loss of strength in a fire if unprotected.

Design

I-beams are widely used in the construction industry and are available in a variety of standard sizes. Tables are available to allow easy selection of a suitable steel I-beam size for a given applied load. I-beams may be used both as beams and as columns.

I-beams may be used both on their own, or acting compositely with another material, typically concrete. Design may be governed by any of the following criteria:

- deflection: the stiffness of the I-beam will be chosen to minimize deformation
- vibration: the stiffness and mass are chosen to prevent unacceptable vibrations, particularly in settings sensitive to vibrations, such as offices and libraries
- bending failure by yielding: where the stress in the cross section exceeds the yield stress
- bending failure by lateral torsional buckling: where a flange in compression tends to buckle sideways or the entire cross-section buckles torsionally

- bending failure by local buckling: where the flange or web is so slender as to buckle locally
- local yield: caused by concentrated loads, such as at the beam's point of support
- shear failure: where the web fails. Slender webs will fail by buckling, rippling in a phenomenon termed tension field action, but shear failure is also resisted by the stiffness of the flanges
- buckling or yielding of components: for example, of stiffeners used to provide stability to the I-beam's web.

Design for Bending

A beam under bending sees high stresses along the axial fibres that are farthest from the neutral axis. To prevent failure, most of the material in the beam must be located in these regions. Comparatively little material is needed in the area close to the neutral axis. This observation is the basis of the I-beam cross-section; the neutral axis runs along the centre of the web which can be relatively thin and most of the material can be concentrated in the flanges.

The ideal beam is the one with the least cross-sectional area (and hence requiring the least material) needed to achieve a given section modulus. Since the section modulus depends on the value of the moment of inertia, an efficient beam must have most of its material located as far from the neutral axis as possible. The farther a given amount of material is from the neutral axis, the larger is the section modulus and hence a larger bending moment can be resisted.

When designing a symmetric I-beam to resist stresses due to bending the usual starting point is the required section modulus. If the allowable stress is σ_{max} and the maximum expected bending moment is M_{max}, then the required section modulus is given by

$$S = \frac{M_{max}}{\sigma_{max}} = \frac{I}{c}$$

where I is the moment of inertia of the beam cross-section and c is the distance of the top of the beam from the neutral axis. For a beam of cross-sectional area a and height h, the ideal cross-section would have half the area at a distance $h/2$ above the cross-section and the other half at a distance $h/2$ below the cross-section For this cross-section

$$I = \frac{ah^2}{4}\ ;\ S = 0.5ah\,.$$

However, these ideal conditions can never be achieved because material is needed in the web for physical reasons, including to resist buckling. For wide-flange beams, the section modulus is approximately

$$S \approx 0.35ah$$

which is superior to that achieved by rectangular beams and circular beams.

Issues

Though I-beams are excellent for unidirectional bending in a plane parallel to the web, they do not perform as well in bidirectional bending. These beams also show little resistance to twisting and undergo sectional warping under torsional loading. For torsion dominated problems, box beams and other types of stiff sections are used in preference to the I-beam.

Wide-Flange Steel Materials and Rolling Processes (U.S.)

In the United States, the most commonly mentioned I-beam is the wide-flange (W) shape. These beams have flanges in which the planes are nearly parallel. Other I-beams include American Standard (designated S) shapes, in which flange surfaces are not parallel, and H-piles (designated HP), which are typically used as pile foundations. Wide-flange shapes are available in grade ASTM A992, which has generally replaced the older ASTM grades A572 and A36. Ranges of yield strength:

- A36: 36,000 psi (250 MPa)
- A572: 42,000–60,000 psi (290–410 MPa), but 50,000 psi (340 MPa) is the most common
- A588: Similar to A572
- A992: 50,000–65,000 psi (340–450 MPa)

Like most steel products, I-beams often contain some recycled content. The American Institute of Steel Construction ("AISC") publishes the "Steel Construction Manual" for designing structures of various shapes. It documents the common approaches, ASD and LRFD, (as of 13th ed.) to creating such designs.

Standards

The following standards define the shape and tolerances of I-beam steel sections:

Euronorms

- EN 10024, Hot rolled taper flange I sections - Tolerances on shape and dimensions.

- EN 10034, Structural steel I and H sections - Tolerances on shape and dimensions.
- EN 10162, Cold rolled steel sections - Technical delivery conditions - Dimensional and cross-sectional tolerances

Other

- DIN 1025-5
- ASTM A6, American Standard Beams
- BS 4-1
- IS 808 - Dimensions hot rolled steel beam, column, channel and angle sections

Designation and Terminology

- In the United States, steel I-beams are commonly specified using the depth and weight of the beam. For example, a "W10x22" beam is approximately 10 inches (25 cm) in depth (nominal height of the I-beam from the outer face of one flange to the outer face of the other flange) and weighs approximately 22 lb/ft (33 kg/m). It should be noted that wide flange section often vary from their nominal depth. In the case of the W14 series, they may be as deep as 22.84 inches (58.0 cm).
- In Canada, steel I-beams are now commonly specified using the depth and weight of the beam in metric terms. For example, a "W250x33" beam is approximately 250 millimetres (9.8 in) in depth (height of the I-beam from the outer face of one flange to the outer face of the other flange) and weighs approximately 33 kg/m (22 lb/ft). I-beams are still available in U.S. sizes from many Canadian manufacturers.
- In IndiaI-beams are designated as ISMB, ISJB, ISLB, ISWB. ISMB: Indian Standard Medium Weight Beam, ISJB: Indian Standard Junior Beams, ISLB: Indian Standard Light Weight Beams, and ISWB: Indian Standard Wide Flange Beams. Beams are designated as per respective abbreviated reference followed by the depth of section, such as for example "ISMB 450", where 450 is the depth of section in millimetres (mm). The dimensions of these beams are classified as per IS:808 (as per BIS).
- In the United Kingdom, these steel sections are commonly specified with a code consisting of the major dimension (usually the depth)-x-the minor dimension-x-the mass per metre-ending with the section type, all measurements being metric. Therefore a 152x152x23UC would be a column section (UC = universal

column) of approximately 152 millimetres (6.0 in) depth 152 mm width and weighing 23 kilograms per metre (46 lb/yd) of length.

- In Australia, these steel sections are commonly referred to as Universal Beams (UB) or Columns (UC). The designation for each is given as the height of the beam, the type (beam or column) and then the unit metre rate (e.g., a 460UB67 is a 460 millimetres (18.1 in) deep universal beam that weighs 67 kilograms per metre (135 lb/yd)).

Cellular Beams

Cellular beams are the modern version of the traditional "castellated" beam which results in a beam approximately 40–60% deeper than its parent section. The exact finished depth, cell diameter and cell spacing are flexible. A cellular beam is up to 1.5 times stronger than its parent section and is therefore utilised to create efficient large span constructions.

Column

Column or pillar in architecture and structural engineering is a structural element that transmits, through compression, the weight of the structure above to other structural elements below. In other words, a column is a compression member. The term column applies especially to a large round support with a capital and base and made of stone, or appearing to be so. A small wooden or metal support is typically called a post, and supports with a rectangular or other non-round section are usually called piers. For the purpose of wind or earthquake engineering, columns may be designed to resist lateral forces. Other compression members are often termed "columns" because of the similar stress conditions. Columns are frequently used to support beams or arches on which the upper parts of walls or ceilings rest. In architecture, "column" refers to such a structural element that also has certain proportional and decorative features. A column might also be a decorative element not needed for structural purposes; many columns are "engaged", that is to say form part of a wall.

History

All significant Iron Age civilizations of the Near East and Mediterranean made some use of columns. In Ancient Egyptian architecture as early as 2600 BC the architect Imhotep made use of stone columns whose surface was carved to reflect the organic form of bundled reeds; in later Egyptian architecture faceted cylinders were also common.

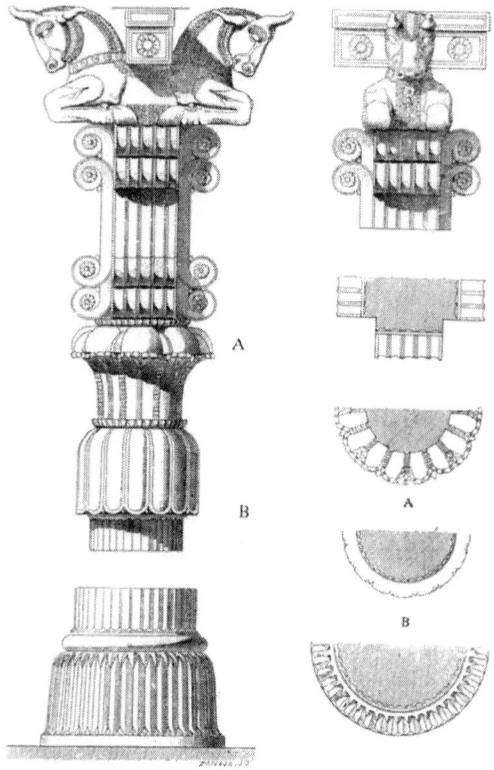

***Figure:** Plan, Front view and side view of a typical Persepolis column, of Persia (Iran).*

Some of the most elaborate columns in the ancient world were those of the Persians, especially the massive stone columns erected in Persepolis. They included double-bull structures in their capitals. The Hall of Hundred Columns at Persepolis, measuring 70 × 70 metres, was built by the Achaemenid king Darius I (524–486 BC). Many of the ancient Persian columns are standing, some being more than 30 metres tall.

The Egyptians, Persians and other civilizations mostly used columns for the practical purpose of holding up the roof inside a building, preferring outside walls to be decorated with reliefs or painting, but the Ancient Greeks, followed by the Romans, loved to use them on the outside as well, and the extensive use of columns on the interior and exterior of buildings is one of the most characteristic features of classical architecture, in buildings like the Parthenon. The Greeks developed the classical orders of architecture, which are most easily distinguished by the form of the column and its various elements.

Their Doric, Ionic, and Corinthian orders were expanded by the Romans to include the Tuscan and Composite orders.

Columns, or at least large structural exterior ones, became much less significant in the architecture of the Middle Ages, and the classical forms were abandoned in both Byzantine architecture and the Romanesque and Gothic architecture or Europe in favour of more flexible forms, with capitals often using various types of foliage decoration, and in the West scenes with figures carved in relief. Renaissance architecture was keen to revive the classical vocabulary and styles, and the informed use and variation of the classical orders remained fundamental to the training of architects throughout Baroque, Rococo and Neo-classical architecture.

Structure

Early columns were constructed of stone, some out of a single piece of stone. Monolithic columns are among the heaviest stones used in architecture. Other stone columns are created out of multiple sections of stone, mortared or dry-fit together. In many classical sites, sectioned columns were carved with a centre hole or depression so that they could be pegged together, using stone or metal pins. The design of most classical columns incorporates entasis (the inclusion of a slight outward curve in the sides) plus a reduction in diameter along the height of the column, so that the top is as little as 83% of the bottom diameter. This reduction mimics the parallax effects which the eye expects to see, and tends to make columns look taller and straighter than they are while entasis adds to that effect.

Nomenclature

Classical columns have a bottom or plinth that rests on the stylobate or foundation. At the top a capital (also known as a chapiter) or finial decorate the column, depending on whether it is attached to a structure or free-standing.

Modern columns are constructed out of steel, poured or precast concrete, or brick. They may then be clad in an architectural covering (or veneer), or left bare. In modern terms the impost (or pier) is the topmost member of a column. The bottom-most part of the arch, called the springing, rests on the impost.

Equilibrium, Instability, and Loads

Mechanical Failure Modes: As the axial load on a perfectly straight slender column with elastic material properties is increased in magnitude, this ideal column passes through three states: stable equilibrium, neutral equilibrium, and instability. The straight column under load is in stable equilibrium if a lateral force, applied between the two ends of the column, produces a small lateral deflection which disappears and the column returns to its straight form when the lateral force is removed. If the column load is gradually increased, a condition is reached in which the straight form of equilibrium becomes so-called neutral equilibrium, and a small lateral force will produce a deflection that does not disappear and the column remains in this slightly bent form when the lateral force is removed. The load at which neutral equilibrium of a column is reached is called the critical or buckling load. The state of instability is reached when a slight increase of the column load causes uncontrollably growing lateral deflections leading to complete collapse.

For an axially loaded straight column with any end support conditions, the equation of static equilibrium, in the form of a differential equation, can be solved for the deflected shape and critical load of the column. With hinged, fixed or free end support conditions the deflected shape in neutral equilibrium of an initially straight column with uniform cross section throughout its length always follows a partial or composite sinusoidal curve shape, and the critical load is given by

$$f_{cr} \equiv \frac{\pi^2 E I_{min}}{L^2} \tag{1}$$

where E = elastic modulus of the material, I_{min} = the minimal moment of inertia of the cross section, and L = actual length of the column between its two end supports. A variant of (1) is given by

$$f_{cr} \equiv \frac{\pi^2 E_T}{\left(\frac{KL}{r}\right)^2} \tag{2}$$

where r = radius of gyration of [column]cross-section which is equal to the square root of (I/A), K = ratio of the longest half sine wave to the actual column length, and KL = effective length (length of an equivalent hinged-hinged column). From Equation (2) it can be noted that the buckling strength of a column is inversely proportional to the square of its length.

When the critical stress, F_{cr} ($F_{cr} = P_{cr}/A$, where A = cross-sectional area of the column), is greater than the proportional limit of the material, the column is experiencing inelastic buckling. Since at this stress the slope of the material's stress-strain curve, E_t (called the tangent modulus), is smaller than that below the proportional limit, the critical load at inelastic buckling is reduced.

More complex formulas and procedures apply for such cases, but in its simplest form the critical buckling load formula is given as Equation (3),

$$f_{cr} \equiv F_y - \frac{F_y^2}{4\pi^2 E}\left(\frac{KL}{r^2}\right) \tag{3}$$

where E_t = tangent modulus at the stress F_{cr}

A column with a cross section that lacks symmetry may suffer torsional buckling (sudden twisting) before, or in combination with, lateral buckling. The presence of the twisting deformations renders both theoretical analyses and practical designs rather complex.

Eccentricity of the load, or imperfections such as initial crookedness, decreases column strength. If the axial load on the column is not concentric, that is, its line of action is not precisely coincident with the centroidal axis of the column, the column is characterized as eccentrically loaded. The eccentricity of the load, or an initial curvature, subjects the column to immediate bending. The increased stresses due to the combined axial-plus-flexural stresses result in a reduced load-carrying ability.

Column elements are considered to be massive if minimal side dimension is equal or more than 400 mm. Massive columns have ability to increase concrete strength during long time period (even during exploitation period). Taking into account possible loads onto structure increase in future (and even threat of progressive failure, terroristic attacks, explosions etc.) massive columns have advantage comparing with not ones. A little economy today has no sense as usual for future.

Moreover relatively small sections are not technological for reinforced structures during their production. Balance between economy, mass of structures and so called "sustainable" construction is necessary.

Extensions

When a column is too long to be built or transported in one piece, it has to be extended or spliced at the construction site. A reinforced concrete column is extended by having the steel reinforcing bars protrude a few inches or feet above the top of the concrete, then placing the next level of reinforcing bars to overlap, and pouring the concrete of the next level. A steel column is extended by welding or bolting splice plates on the flanges and webs or walls of the columns to provide a few inches or feet of load transfer from the upper to the lower column section. A timber column is usually extended by the use of a steel tube or wrapped-around sheet-metal plate bolted onto the two connecting timber sections.

when e<12 then it is known as short column

Foundations

A column that carries the load down to a foundation must have means to transfer the load without overstressing the foundation material. Reinforced concrete and masonry columns are generally built directly on top of concrete foundations. A steel column, when seated on a concrete foundation, must have a base plate to spread the load over a larger area and thereby reduce the bearing pressure. The base plate is a thick rectangular steel plate usually welded to the bottom end of the column.

Classical Orders

The Roman author Vitruvius, relying on the writings (now lost) of Greek authors, tells us that the ancient Greeks believed that their Doric order developed from techniques for building in wood in which the earlier smoothed tree trunk was replaced by a stone cylinder.

Doric Order

The Doric order is the oldest and simplest of the classical orders. It is composed of a vertical cylinder that is wider at the bottom. It generally has neither a base nor a detailed capital. It is instead often topped with an inverted frustum of a shallow cone or a cylindrical band of carvings. It is often referred to as the masculine order because it is represented in the bottom level of the Colosseum and the Parthenon, and was therefore considered to be able to hold more weight. The height-to-thickness ratio is about 8:1. The shaft of a Doric Column is always fluted.

The Greek Doric, developed in the western Dorian region of Greece, is the heaviest and most massive of the orders. It rises from the stylobate without any base; it is from four to six times as tall as its diameter; it has twenty broad flutes; the capital consists simply of a banded necking swelling out into a smooth echinus, which carries a flat square abacus; the Doric entablature is also the heaviest, being about one-fourth the height column. The Greek Doric order was not used after c. 100 B.C. until its "rediscovery" in the mid-eighteenth century.

Tuscan Order

The Tuscan order, also known as Roman Doric, is also a simple design, the base and capital both being series of cylindrical disks of alternating diameter. The shaft is almost never fluted. The proportions vary, but are generally similar to Doric columns. Height to width ratio is about 7:1.

Ionic Order

The Ionic column is considerably more complex than the Doric or Tuscan. It usually has a base and the shaft is often fluted (it has grooves carved up its length). On the top is a capital in the characteristic shape of a scroll, called a volute, or scroll, at the four corners. The height-to-thickness ratio is around 9:1. Due to the more refined proportions and scroll capitals, the Ionic column is sometimes associated with academic buildings. Ionic style columns were used on the second level of the Colosseum.

Corinthian Order

The Corinthian order is named for the Greek city-state of Corinth, to which it was connected in the period. However, according to the architectural historian Vitruvius, the column was created by the sculptor Callimachus, probably an Athenian, who drew acanthus leaves growing around a votive basket. In fact, the oldest known Corinthian capital was found in Bassae, dated at 427 BC. It is sometimes called the feminine order because it is on the top level of the Colosseum and holding up the least weight, and also has the slenderest ratio of thickness to height. Height to width ratio is about 10:1.

Composite Order

The Composite order draws its name from the capital being a composite of the Ionic and Corinthian capitals. The acanthus of the Corinthian column already has a scroll-like element, so the distinction is sometimes subtle. Generally the Composite is similar to the Corinthian in proportion and employment, often in the upper tiers of colonnades. Height to width ratio is about 11:1 or 12:1.

Solomonic

A Solomonic column, sometimes called "barley sugar", begins on a base and ends in a capital, which may be of any order, but the shaft twists in a tight spiral, producing a dramatic, serpentine effect of movement. Solomonic columns were developed in the ancient world, but remained rare there. A famous marble set, probably 2nd century, was brought to St Peter's, Rome by Constantine I, and placed round the saint's shrine, and was thus familiar throughout the Middle Ages, by which time they were thought to have been removed from the Temple of Jerusalem. The style was used in bronze by Bernini for his spectacular St. Peter's baldachin, actually a ciborium (which displaced Constantine's columns), and thereafter became very popular with Baroque and Rococo church architects, above all in Latin America, where they were very often used, especially on a small scale, as they are easy to produce in wood by turning on a lathe (hence also the style's popularity for spindles on furniture and stairs).

Joist

In architecture and engineering, a joist is one of the horizontal supporting members that run from wall to wall, wall to beam, or beam to beam to support a ceiling, roof, or floor. It may be made of oriented strand board, plywood, wood, steel, or concrete. Typically, a beam is bigger than a joist, which is often supported by beams and laid out in repetitive patterns.

The wider the span between the supporting structures, the deeper the joist will need to be if it is not to deflect under load. Lateral support also increases its strength. There are approved formulas for calculating the depth required and reducing the depth as needed; however, a rule of thumb for calculating the depth of a wooden floor joist for a residential property is half the span in feet plus two inches; for example, the joist depth required for a 14 foot span is 9 inches. Many steel joist manufacturers supply load tables in order to allow designers to select the proper joist sizes for their projects.

Engineered wood products such as I-joists gain strength from the depth of the floor or the height of each joist, as well as the larger bottom and top chords, as compared to standard dimensional lumber joists. A common saying regarding structural design is that "deeper is cheaper", referring to the more cost-effective design of a given structure by using deeper but more expensive joists, because fewer joists needed and longer spans are achieved, which more than makes up for the added cost of deeper joists.

Bandsill is another term for joist used by construction workers and home inspectors in the southeastern United States.

In traditional timber framing there may be a single set of joists which carry both a floor and ceiling called a single floor (single joist floor, single framed floor) or two sets of joists, one carrying the floor and another carrying the ceiling called a double floor (double framed floor). The term binding joist is sometimes used to describe beams at floor level running perpendicular to the ridge of a gable roof and joined to the intermediate posts. Joists which land on a binding joist are called bridging joists. A large beam in the ceiling of a room carrying joists is a summer beam. A ceiling joist may be installed flush with the bottom of the beam or sometimes below the beam. Joists left exposed and visible from below are called "naked flooring" or "articulated" (a modern U.S. term) and were typically planed smooth (wrought) and sometimes chamfered or beaded. Joists may join to their supporting beams in many ways: joists resting on top of the supporting beams are said to be "lodged"; dropped in using a butt cog joint (a type of lap joint), half-dovetail butt cog, or a half-dovetail lap joint. Joists may also be tenoned in during the raising with a soffit tenon or a tusk tenon (possibly with a housing). Joists can also be joined by being slipped into mortises after the beams are in place such as a chase mortise (pulley mortise), L-mortise, or "short joist". Also, in some Dutch-American work, ground level joists are be placed on a foundation and then a sill placed on top of the joists such as what Jack Sobon called an "inverted sill" or with a "plank sill".

Joists can have different joints on either ends such as being tenoned on one end and lodged on the other end. A reduction in the under-side of cogged joist-ends may be square, sloped or curved. Typically joists do not tie the beams together, but sometimes they are pinned or designed to hold under tension. Joists on the ground floor were sometimes a pole (pole joist, half round joist, log joist. A round timber with one flat surface) and in barns long joists were sometimes supported on a sleeper (a timber not joined to but supporting other beams). Joists left out of an area form an opening called a "well" as in a stairwell or chimney-well.

The joists forming the well are the heading joist (header) and trimming joist (trimmer). Trimmers take the name of the feature such as hearth trimmer, stair trimer, etc. Shortened joists are said to be crippled. The term rim joist is rare before the 1940s in America, it forms the edge of a floor. The outermost joist in half timber construction may be of a more durable species than the interior joists. In a barn, loose poles above the drive floor are called a scaffold. Between the joists, the area called a joist-bay, and above the ceiling in some old houses is material called pugging which was used to deaden sound, insulate, and resist the spread of fire.

Deck (Building)

In architecture, a deck is a flat surface capable of supporting weight, similar to a floor, but typically constructed outdoors, often elevated from the ground, and usually connected to a building. The term is a generalization of decks as found on ships.

Functions and Materials

Wood or timber "decking" can be used in a number of ways: as part of garden landscaping, to extend living areas of houses, and as an alternative to stone based features such as patios. Decks are made from treated lumber, composite lumber, composite material, Aluminium, Western red cedar, teak, mahogany, ipê and other hardwoods and recycled planks made from high-density polyethylene (HDPE), polystyrene (PS) and PET plastic as well as mixed plastics and wood fibre (often called "composite" lumber). Artificial decking products are often called "wood-plastic composites". These days, WPC's have more widely known by different brands like Trex, Azek, Ecornboard etc.

Historically, the softwoods used for decking were logged from old growth forests. These include Atlantic white cedar, redwood and Western red cedar (redcedar). Atlantic City built the first coastal boardwalk in the United States, originally constructed of Atlantic white cedar. However, it was not long before the commercial logging of this tree and clearing of cedar swamps in New Jersey caused a decline in the availability of decking. Atlantic City and New York City both switched to Western red cedar. By the 1960s, Western red cedar from the US was declining due to over-logging. More expensive Western red cedar was available from western Canada (British Columbia) but by then, pressure treated pine had become available.

But even with chemical treatments (such as chromated copper arsenate or CCA), pine decking is not as durable as cedars in an outdoor environment. Thus, many municipalities and homeowners are turning to hardwoods. Decks are often built from pressure treated wood. Pressure treated wood is long lasting and holds up to wet and icey weather conditions. Pressure treated wood however is treated with chemicals which have been known to be toxic. Slivers received from pressure treated wood most generally become infected. Pressure treated saw dust also contains toxins such as strychnine, also often used as rat poison. These toxins, when inhaled, can require hospitalization for both acute and chronic exposures.

Generally, hardwoods used for decking come from tropical forests. Much of the logging taking place to produce these woods, especially teak, mahogany and ipê, is occurring illegally, as outlined in numerous reports by environmental organisations such as Greenpeace, Friends of the Earth and Rainforest Relief. US tropical wood imports are rising, partly due to the demand for decking.

Due to environmental concerns, composite decking (a mixture of two materials, typically wood pulp and recycled material such as plastic bottles or plastic bags) have appeared on the market. Proponents of composite decking have touted this as a much needed development as this helps to curb logging of trees for new decks. However composite decking has been found to contain harmful chemicals, cannot be refurbished, and despite claims from decking companies, the composite deck still attracts molding.

Residential decks may contain spaces for BBQing, dining and seating. BBQ areas ideally should be situated near the patio door while out of the way from general foot traffic. Dining spaces will include patio tables, for a typical 6 person outdoor patio table building an area of 12' x 16' is ideal. If deck space is available, homeowners may choose to include a seating area for outdoor couches and benches.

Construction

The deck of a house is generally a wooden platform built above the ground and connected to the main building. It is generally enclosed by a railing for safety. Access may be from the house through doors and from the ground via a stairway. Residential decks can be constructed over steep areas or rough ground that is otherwise unusable. Decks can also be covered by a canopy or pergola to control sunlight. Deck designs can be found in numerous books, do-it-yourself magazines and web sites, and from the USDA.

Larger buildings may also have decks on the upper floors of the building which may be open to the public as observation decks or a Skyrise greenery. A deck is also the surface used to construct a boardwalk over sand on barrier islands.

Laying deck or throwing deck refers to the act of placing and bolting down cold-formed steel beneath roofing and concrete floors. This is usually done by an ironworker, sometimes in conjunction with a cement mason or carpenter. It regarded as one of the most physically demanding jobs in the iron working industry.

In the UK the various ban on smoking in public buildings was expected lead to an increase in the use of timber decking for outdoor spaces where smokers can gather.

Roof Deck

The roof deck is the roofing material layer between the primary structural components (trusses&joists) and either insulative layers or weatherproofing layers in a typical roof system. Usually the roof deck acts as a unifying structural diaphragm by tying all the structural components together. There are many types of roof deck: plywood or OSB sheathing, wood tongue and groove, corrugated metal, grancrete encapsulated polystyrene, reinforced concrete, and double tee.

Framing (Construction)

Framing, in construction known as light-frame construction, is a building technique based around vertical structural members, usually called studs, which provide a stable frame to which interior and exterior wall coverings are attached, and covered by a roof made of horizontal ceiling joists and sloping rafters (or pre-fabricated roof trusses).

Modern light-frame structures usually gain strength from rigid panels (plywood and other plywood-like composites such as oriented strand board (OSB) used to form all or part of wall sections) but until recently carpenters employed various forms of diagonal bracing (called *wind braces*) to stabilize walls. Diagonal bracing remains a vital interior part of many roof systems, and in-wall wind braces are required by building codes in many municipalities or by individual state laws in the United States.

Light frame construction using standardized dimensional lumber has become the dominant construction method in North America and Australia because of its economy. Use of minimal structural materials allows builders to enclose a large area with minimal cost, while achieving a wide variety of architectural styles. The ubiquitous platform framing and the older balloon framing are the two different light frame construction systems used in North America.

Walls

Wall framing in house construction includes the vertical and horizontal members of exterior walls and interior partitions, both of bearing walls and non-bearing walls. These *stick* members, referred to as studs, wall plates and lintels (*headers*), serve as a nailing base for all covering material and support the upper floor platforms, which provide the lateral strength along a wall. The platforms may be the boxed structure of a ceiling and roof, or the ceiling and floorjoists of the story above. The technique is variously referred to colloquially in the building trades as *stick and frame, stick and platform,* or *stick and box* as the sticks (studs) give the structure its vertical support, and the box-shaped floor sections with joists contained within length-long post and lintels (more commonly called *headers*), support the weight of whatever is above, including the next wall up and the roof above the top story. The platform also provides the lateral support against wind and holds the stick walls true and square. Any lower platform supports the weight of the platforms and walls above the level of its component headers and joists.

Framing lumber should be grade-stamped, and have a moisture content not exceeding 19%.

There are three historically common methods of framing a house.

- Post and beam, which is now used predominately in barn construction.
- Balloon framing using a technique suspending floors from the walls was common until the late 1940s, but since that time, platform framing has become the predominant form of house construction.
- Platform framing often forms wall sections horizontally on the sub-floor prior to erection, easing positioning of studs and increasing accuracy while cutting the necessary manpower. The top and bottom plates are end-nailed to each stud with two nails at least 3.25 in (83 mm) in length (***16d*** or ***16** penny* nails). Studs are at least doubled (creating posts) at openings, the jack stud being cut to receive the lintels (headers) that are placed and end-nailed through the outer studs.

Wall sheathing, usually a plywood or other laminate, is usually applied to the framing prior to erection, thus eliminating the need to scaffold, and again increasing speed and cutting manpower needs and expenses. Some types of exterior sheathing, such as asphalt-impregnated fibreboard, plywood, oriented strand board and waferboard, will provide adequate bracing to resist lateral loads and keep the wall square. (Construction codes in most jurisdictions require a stiff plywood sheathing.) Others, such as rigid glass-fibre, asphalt-coated fibreboard, polystyrene or polyurethane board, will not. In this latter case, the wall should be reinforced with a diagonal wood or metal bracing inset into the studs. In jurisdictions subject to strong wind storms (hurricane countries, tornado alleys) local codes or state law will generally require both the diagonal wind braces and the stiff exterior sheathing regardless of the type and kind of outer weather resistant coverings.

Corners

A multiple-stud post made up of at least three studs, or the equivalent, is generally used at exterior corners and intersections to secure a good tie between adjoining walls and to provide nailing support for the interior finish and exterior sheathing. Corners and intersections, however, must be framed with at least two studs. Nailing support for the edges of the ceiling is required at the junction of the wall and ceiling where partitions run parallel to the ceiling joists. This material is commonly referred to as 'dead wood' or backing.

Exterior Wall Studs

Wall framing in house construction includes the vertical and horizontal members of exterior walls and interior partitions. These members, referred to as studs, wall plates and lintels, serve as a nailing base for all covering material and support the upper floors, ceiling and roof.

Exterior wall studs are the vertical members to which the wall sheathing and cladding are attached. They are supported on a bottom plate or foundation sill and in turn support the top plate. Studs usually consist of 1.5 in × 3.5 in (38 mm × 89 mm) or 1.5 in × 5.5 in (38 mm × 140 mm) lumber and are commonly spaced at 16 in (410 mm) on centre. This spacing may be changed to 12 in (300 mm) or 24 in (610 mm) on centre depending on the load and the limitations imposed by the type and thickness of the wall covering used. Wider 1.5 in × 5.5 in (38 mm × 140 mm) studs may be used to provide space for more insulation. Insulation beyond that which can be accommodated within a 3.5 in (89 mm) stud space can also be provided by other means, such as rigid or semi-rigid insulation or batts between 1.5 in × 1.5 in (38 mm × 38 mm) horizontal furring strips, or rigid or semi-rigid insulation sheathing to the outside of the studs. The studs are attached to horizontal top and bottom wall plates of 1.5 in (38 mm) lumber that are the same width as the studs.

Interior Partitions

Interior partitions supporting floor, ceiling or roof loads are called loadbearing walls; others are called non-loadbearing or simply partitions. Interior loadbearing walls are framed in the same way as exterior walls. Studs are usually 1.5 in × 3.5 in (38 mm × 89 mm) lumber spaced at 16 in (410 mm) on centre. This spacing may be changed to 12 in (300 mm) or 24 in (610 mm) depending on the loads supported and the type and thickness of the wall finish used.

Partitions can be built with 1.5 in × 2.5 in (38 mm × 64 mm) or 1.5 in × 3.5 in (38 mm × 89 mm) studs spaced at 16 or 24 in (400 or 600 mm) on centre depending on the type and thickness of the wall finish used. Where a partition does not contain a swinging door, 1.5 in × 3.5 in (38 mm × 89 mm) studs at 16 in (410 mm) on centre are sometimes used with the wide face of the stud parallel to the wall. This is usually done only for partitions enclosing clothes closets or cupboards to save space. Since there is no vertical load to be supported by partitions, single studs may be used at door openings. The top of the opening may be bridged with a single piece of 1.5 in (38 mm) lumber the same width as the studs. These members provide a nailing support for wall finish, door frames and trim.

Lintels (Headers)

Lintels (or, headers) are the horizontal members placed over window, door and other openings to carry loads to the adjoining studs. Lintels are usually constructed of two pieces of 2 in (nominal) (38 mm) lumber separated with spacers to the width of the studs and nailed together to form a single unit. The preferable spacer material is rigid insulation. The depth of a lintel is determined by the width of the opening and vertical loads supported.

Wall Sections

The complete wall sections are then raised and put in place, temporary braces added and the bottom plates nailed through the subfloor to the floor framing members. The braces should have their larger dimension on the vertical and should permit adjustment of the vertical position of the wall.

Once the assembled sections are plumbed, they are nailed together at the corners and intersections. A strip of polyethylene is often placed between the interior walls and the exterior wall, and above the first top plate of interior walls before the second top plate is applied to attain continuity of the air barrier when polyethylene is serving this function.

A second top plate, with joints offset at least one stud space away from the joints in the plate beneath, is then added. This second top plate usually laps the first plate at the corners and partition intersections and, when nailed in place, provides an additional tie to the framed walls. Where the second top plate does not lap the plate immediately underneath at corner and partition intersections, these may be tied with 0.036 in (0.91 mm) galvanized steel plates at least 3 in (76 mm) wide and 6 in (150 mm) long, nailed with at least three 2.5 in (64 mm) nails to each wall.

Balloon Framing

Balloon framing is a method of wood construction – also known as "Chicago construction" in the 19th century – used primarily in Scandinavia, Canada and the United States (up until the mid-1950s). It utilises long continuous framing members (studs) that run from the sill plate to the top plate, with intermediate floor structures let into and nailed to them. Here the heights of window sills, headers and next floor height would be marked out on the studs with a storey pole. Once popular when long lumber was plentiful, balloon framing has been largely replaced by *platform framing*. It is not certain who introduced balloon framing in the United States. However, the first building using balloon framing was probably a warehouse constructed in 1832 in Chicago, Illinois, by George Washington Snow. The following year, Augustine Taylor (1796–1891) constructed St. Mary's Catholic Church in Chicago using the balloon framing method.

In the 1830s, Hoosier Solon Robinson published articles about a revolutionary new framing system, called "balloon framing" by later builders. Robinson's system called for standard 2x4 lumber, nailed together to form a sturdy, light skeleton. Builders were reluctant to adopt the new technology, however, by the 1880s, some form of 2x4 framing was standard.

Alternatively, the balloon frame has been shown to have been introduced in Missouri as much as fifty years earlier.

The name comes from a French Missouri type of construction, *maison en boulin, boulin* being a French term for a horizontal scaffolding support. Historians have also fabricated the following story: As Taylor was constructing his first such building, St. Mary's Church, in 1833, skilled carpenters looked on at the comparatively thin framing members, all held together with nails, and declared this method of construction to be no more substantial than a balloon. It would surely blow over in the next wind! Though the criticism proved baseless, the name stuck.

Although lumber was plentiful in 19th-century America, skilled labour was not. The advent of cheap machine-made nails, along with water-powered sawmills in the early 19th century made balloon framing highly attractive, because it did not require highly-skilled carpenters, as did the dovetail joints, mortises and tenons required by post-and-beam construction. For the first time, any farmer could build his own buildings without a time-consuming learning curve.

Figure: *A very unusual example of balloon framing: The Jim Kaney Round Barn, Adeline, Illinois, U.S.A.*

It has been said that balloon framing populated the western United States and the western provinces of Canada. Without it, western boomtowns certainly could not have blossomed overnight. It is also a fair certainty that, by radically reducing construction costs, balloon framing improved the shelter options of poorer North Americans. For example, many 19th-century New England working neighbourhoods consist of balloon-constructed three-story apartment buildings referred to as triple deckers. The main difference between platform and balloon framing is at the floor lines. The balloon wall studs extend from the sill of the first story all the way to the top plate or end rafter of the second story. The platform-framed wall, on the other hand, is independent for each floor.

Balloon framing has several disadvantages as a construction method:

1. The creation of a path for fire to readily travel from floor to floor. This is mitigated with the use of firestops at each floor level.
2. The lack of a working platform for work on upper floors. Whereas workers can readily reach the top of the walls being

erected with platform framing, balloon construction requires scaffolding to reach the tops of the walls (which are often two or three stories above the working platform).

3. The requirement for long framing members.
4. In certain larger buildings, a noticeable down-slope of floors towards central walls, caused by the differential shrinkage of the wood-framing members at the perimeter versus central walls. Larger balloon-framed buildings will have central bearing walls which are actually platform framed and thus will have horizontal sill and top plates at each floor level, plus the intervening floor joists, at these central walls. Wood will shrink much more across its grain than along the grain. Therefore, the cumulative shrinkage in the centre of such a building is considerably more than the shrinkage at the perimeter where there are many fewer horizontal members. This problem, unlike the first three, takes time to develop and become noticeable.
5. Present-day balloon framing buildings often have higher heating costs, due to the lack of insulation separating a room from its exterior walls. However, this can be remedied through the addition of insulation, as with any other framed building.

Since steel is generally more fire-resistant than wood, and steel framing members can be made to arbitrary lengths, balloon framing is growing in popularity again in light gauge steel stud construction. Balloon framing provides a more direct load path down to the foundation. Additionally, balloon framing allows more flexibility for tradesmen in that it is significantly easier to pull wire, piping and ducting without having to bore through or work around framing members.

Platform Framing

In Canada and the United States, the most common method of light-frame construction for houses and small apartment buildings as well as other small commercial buildings is *platform framing*. In builder parlance, platform framing might also nowadays be called (only partly correctly) '*stick framing*' or '*stick construction*' as each element is built up stick by stick, which was also true in the other stick framing method, in the obsolete and labour intensive, but previously fashionable, balloon framing method, wherein the outside walls were erected, headers hung, then floor joists were inserted into a box made of walls. In contrast, in platform framing a floor box and joists making up the platform is built and placed on a supporting under structure (Sill plates, headers, or beams) where it sits flat and gets fastened down against wind lifting with galvanized metal tie straps. Once the boxed floor platform is squared, leveled and fastened then subfloor, walls, ceilings, and roof are built onto and above that initial platform, which can be repeated floor by floor, 'without the slow downs and dangers of fastening and leveling rough-sawn joists of a new floor together to the walls from ladders extending one or even two stories up.

Generally, the flooring ('platform') is constructed then the walls built on top of that layer, then another atop that, and so forth making for quick efficient labour saving construction methodologies and those have quickened further as technologies such as joist hangers have been developed to speed and enhance the technology. The methods and techniques have become so common and pervasive that even Skyscrapers use a modified form of platform framing techniques and indeed the same tools and technologies once construction builds the initial structural skeleton. Once the platform floor is laid down, the builder's crew can with chalk line, rule and pencil directly transfer an outline of the exterior and interior walls, their openings and relative locations with ease and precision from the plans or builders blue prints.

As the survey group lays down the notations and chalk lines, a carpenter crew can follow behind and lay down 2x4 'bottom plates' and tack them to the floor box. The topmost wall plates are cut only to the outside dimensions of the walls. Butting two other two by fours against these cut to size and fastened bottom plate allows the crew to rule across all three with square and lay out studs, cripple studs, and openings for that particular wall. The two loose studs are then quickly flipped on edge after openings are cut in, and studs added on the marks with quick reliable end nailing through the respective top and bottom plates. A few minutes later the whole wall section can be levered up and aligned in place and braced for later application of the top plates and adjoining walls.

The method provides builders options and flexibility such as when and where there is a floor-level opening (doorway) the next wall section can be aligned and fastened in place separately with the top plate added then used then a lintel and cripple studding added, or the entire wall could have been cut and joined at the top all along and lifted up as one entity. In the end, the outside walls are plumbed and fastened together with 'ell-configured reinforced corners' that provide nailing wood in the interior angles and strength to the building forming in effect wide posts at each corner and fastened lastly by overlapped top plates which stagger their joints from the ones capping each plate by which the studs are end nailed together. Each wall from top to bottom ends up with a doubled plate, studs, and a doubled plate, where structurally the doubled plates spread the weight of the roof and loading across the studs of the wall, ultimately to the foundation.

Overall, the framed structure sits (most commonly) atop a concrete foundation on *pressure treated wood 'sill', or 'beam'*. When on concrete, the sill plate is anchored, usually with (embedded) 'J' bolts into the concrete substrate of the foundation wall. Generally these plates must be pressure treated to keep from rotting from condensing moisture. By various standards the bottom of the sill plate is located a minimum 6 inches (150 mm) above the *finished grade* (the surrounding ground) per standard builders practices, and frequently more dependent upon building codes of the relevant jurisdiction's local building codes. In North America, building codes may differ not only state to state, but town to town, the tighter specification applying at all times. This distance, together with roofing overhangs, and other system factors, is most often selected both to prevent the sill-plate from rotting (due to the invasion of splashed water) as well as providing a termite barrier. The latter is particularly (more or less) important than anti-rotting considerations depending upon the geographical location.

Alternatively, the room, room extension, deck or even a house can be built above concrete columns U.S. builders call piers some others call pilasters, another of many term misuses common to building trade parlance. In such cases, the pier (column) is usually required to rest on bed rock or extend well below the zone of average freezing soil depth (the same as a foundation) locally, and frequently is required to also have flared out or mushroomed bottom of greater surface than that the pier top (these are called 'big foots' in the building trade, and building suppliers carry PVC molds to conserve concrete which allow a builder to satisfy area requirements and the building codes). Rigid pressure treated 'beams' (usually doubled or tripled up wider types of 2x boards) are attached to the piers using galvanized metal brackets and serve the same function as sills in foundation supported framing.

The floors, walls and roof of a framed structure are created by assembling (using nails) consistently sized framing elements of dimensional lumber (e.g. 2×4s) at regular spacings (typically divisions of 4 and 8 feet, or such as 12, 16, 19.2, or 24 inches on centre). The empty space formed between elements is called a stud bay in the wall and a joist bay in the floor or ceiling. The floors, walls and roof are typically made torsionally stable with the installation of a plywood or composite wood skin referred to as sheathing. Sheathing has very specific requirements (such as thickness and spacing of nailing). These measures allow a known amount of shear force to be resisted by the elements. Spacing the framing members properly usually allows them to align with the edges of standard sheathing. In the past, tongue and groove planks installed diagonally were used as sheathing. Occasionally, wooden or galvanized steelbraces are used instead of sheathing. There are also engineered wood panels made for shear and bracing.

The floor, or the platform in this framing type's name, is made up of joists (usually 2x6, 2×8, 2×10 or 2×12 depending on the span, on edge thus the wider joist supporting weight for a greater distance) that sit on supporting foundation walls, beams, columns or girders within and at right angle to doubled outside members also on edge (the band), forming a box. The outer perimeter is nearly the same (3 inch vs. 3.5 inches) width as the support sill. The joists will generally be installed across the shortest distance of any floor span rectangle. The outer layer of the band will overlap the inner layer with staggered end joints creating a stronger box. If joist hangers are not used, the installation of the outer board in the band is delayed to allow through-nailing directly into the ends of the joists.

The floor joists are spaced at 12 in, 16 in, and 24 in on centre, depending upon the live load needs of the design – the closer the spacing and the wider the floor joist dimension, the less the floor will flex. It is then usually covered with a 3/4-inch tongue-and-groove plywood subfloor. In the century past, 1x planks set at 45-degrees to the joists were used for the first subfloor layer, and a second layer of 1x planks set at 90-degrees to the floor cladding topped that as the second subfloor layer. In that same era, all flooring choices were a very short menu of choices between finished wood types or ceramic tiles versus today's extensive multipage menu of manufactured flooring types. Where the design calls for a framed floor, the resulting platform is where the framer will construct and stand that floor's walls (interior and exterior load bearing walls and space-dividing, non-load bearing partitions). Additional framed floors and their walls may then be erected to a general maximum of four in wood framed construction. There will be no framed floor in the case of a single-level structure with a concrete floor known as a *slab on grade*.

Stairs between floors are framed by installing three 90°-stepped *stringers* attached to wall structures and then placing the horizontal *treads* and vertical *risers* (usually about 14 of each for an 8-ft. ceiling) upon the planes formed by the stringers. A framed roof is an assembly of rafters and wall-ties supported by the top story's walls. Prefabricated and site-built trussed rafters are also used along with the more common stick framing method. *Trusses* are engineered to redistribute tension away from wall-tie members and the ceiling members. The roof members are covered with sheathing or strapping to form the roof deck for the finish roofing material. Floor joists can be engineered lumber (trussed, I-joist, etc.), conserving resources with increased rigidity and value. They are semi-custom manufactured to allow access for runs of plumbing, HVAC, etc. and *some 'common-needs' forms* are pre-manufactured as semi-mass-produced standard products made on a per order basis, like roofing trusses. Such products have a post-order lead time from several weeks to several months.

Double framing is a style of framing used in some areas to reduce heat loss and air infiltration. Two walls are built around the perimeter of the building with a small gap in between. The inner wall carries the structural load of the building and is constructed as described above. The exterior wall is not load bearing and can be constructed using lighter materials. Insulation is installed in the entire space between the outside edge of the exterior wall and the inside edge of the interior wall. The size of the gap depends upon how much insulation is desired. The vapor barrier is installed on the outside of the inner wall, rather than between the studs and drywall of a standard framed structure. This increases its effectiveness as it is not perforated by electrical and plumbing connections.

5

Framing of Stores and Warehouses

Although the larger stores and warehouses, particularly those in large cities, are now generally constructed with iron or steel posts and joists, yet the larger proportion of two and three-story business buildings, lodging houses, etc., are still built with wooden floors and partitions, and will probably continue to be for many years to come. Steel construction is undoubtedly preferable to wood in many ways, and when it can be employed and allow a fair return on the investment, the architect should certainly recommend it to the owner.

There are many ways, however, in which the ordinary wood construction, as found in the smaller cities and in country towns, may be greatly improved and made to answer its purposes nearly as well as steel construction.

One of the greatest defects in the ordinary construction of business blocks, small office buildings, etc., is supporting the floors on woodenpartitions. No building other than a dwelling should have the floors supported in that way.

If the distance between the walls is not more than 24 feet the floor joists should be made of sufficient size to span from wall to wall without assistance from partitions. If the distance between the walls is greater than 25 feet then brick partition walls, or posts and girders should be used for intermediate supports. For large rooms single spans of 28 feet may be used, but this is about the maximum span for which wooden joists can be used with economy.

The principal objections to the use of partitions for supporting the floor joists in this class of buildings are:

1. Inconvenience in changing partitions. Business blocks in particular often require alterations in the partitions to suit the convenience of tenants, and when the partitions support the floors they cannot well be moved.
2. The weight of the joists on the partitions has a tendency to spring the studding and loosen the plastering. The partitions themselves often have insufficient supports, and the ordinary 2-inch cap is too thin to support heavy floors. It will generally be found that where floors are supported on partitions that the plastering on the partitions is badly cracked and the ceiling sags at the centre of the partitition.
3. Less security in case of fire. Stud partitions, being constructed of small timbers, are quickly consumed by fire, and but a few moments are required to weaken the studding sufficiently to cause the floors to fall. Girders, on the contrary, being large, solid timbers, do not burn readily, and will often stand until the fire is extinguished.

Posts and Girders: These should be arranged so that the span of the joists will not exceed 24 feet, and 16 feet gives greater economy. The girders may be of wood or steel; the former will generally be used except where girders of considerable length are required. With a joist span of 16 feet it is not good practice to have a greater span than 14 feet for the girders, and 12 feet is the maximum span permitted in several cities for wooden girders.

The posts may be either of wood, cast iron or steel. In buildings of not more than three stories, having wooden floor beams and girders, iron columns offer no particular advantage over wood, except that they may be made a little smaller. Metal posts, unprotected, will not stand as long in a fire as heavy wooden posts.

Terra-Cotta Floor Arches

Hollow terra-cotta blocks, moulded in the form of a flat arch, are used to a large extent for fireproof floors, and are to be obtained in a variety of patterns and devices. These various patterns may be divided into two principal classes, the side method and the end method.

Side Method

The side method arch, where the blocks are laid with the webs parallel to the beams is the original form of terra-cotta floor arch, while the end method, where the blocks are laid end to end at right angles to the beams, is a later improvement, designed to present the full end section of the material to resist the great thrust of the arches. These side method arches are usually made of dense terra-cotta and may be obtained of various depth from six to fifteen inches, and they should be set with close joints, and be thoroughly cemented together. Specially moulded blocks, called "skewbacks," are made to fit the lower flanges of the beams and project about two inches below the beam, which is covered by a thin strip of tile. The space above the blocks is filled with a cinder concrete, in which bevelled wood strips are embedded for a nailing for the wooden floors.

Side method arches are made to break joints endways, so as to give a bond; and they are usually strong enough for all ordinary floor loads. The joints in the blocks are generally made parallel to the sides of the key block, as this gives a uniform pattern, and so is less expensive than a radial jointing, though the latter would make a stronger arch, but on account of the expense of the different patterns, to make and adjust, it is little used.

End Method

In this method, the blocks are usually made of porous terra-cotta and are set end to end, giving greater resistance to the thrust by forming a series of continuous webs from beam to beam. In this system, the blocks are usually set in continuous lines, not breaking joints. The jointing of these tiles must be done with great care, as the open ends do not give so good a surface for cementing as in the side method, and, the bearing on the beams being given by the thin webs, it is necessary that they should fit perfectly.

This is done to gain the extra strength of the end blocks, and the better bearing of the flat skewbacks, so that the skewbacks are made with many webs and of small sections.

Setting

Floor tiles of either pattern must be set upon plank centers which are hung from the beams, and should be crowned one-quarter of an inch in an arch of six feet. All joints must be close, and made with cement. The centering should be left in place until the cement is thoroughly set, which will require from twelve to thirty-six hours, according to the weather and the nature of the cement, and care must be taken that the freshly laid tiles are not too heavily loaded with materials until they are hard. If the arches are to be plastered upon for a ceiling, they must be kept clean, or bad stains are likely to appear. All holes or irregularities on the under side must be filled with cement mortar to give a proper surface for plastering. Where it is not required to have a flat ceiling, but strength is the main factor to be observed, segmental tile arches are often used. These may be used, with a rise of an inch to the foot, up to twenty feet of span, and are employed to great extent in warehouse construction.

While the nature of floor tiles will not permit of a fine joint being made, they should be laid as closely as possible, especially the key blocks. Joints as great as a half-inch should not be permitted, and the tiles should be set in place by being "shoved" together as in brick laying. Unless the building is closed in, floor tiles cannot be laid in cold climates when there is danger of freezing and thawing, as the joints are liable to be affected to the extent of causing deflection if not more serious trouble.

Floor arches are often tested by applying a heavy roller to the arches after the wood centers have been removed, and by dropping a heavy block of timber upon the arch, a two-inch bed of sand being previously spread over the tiles to prevent mechanical damage.

Roof of Agricultural Building, Pan-American Exposition

This was one of the four largest buildings of the Exposition, being 150 feet wide and 500 feet long. Thus giving a clear central hall 97 feet wide and side aisles each about 17 feet wide, between columns.

The small columns are built up of Norway pine planks laid flat and breaking joint every 3 feet, each having two 5/8-inch staggered bolts 12 inches apart, one through the column on each side of each joint. The edges of the planks are protected by a plank on each side of the column spiked to every plank with sixty-penny spikes 18 inches apart. Each of the large columns has two principal members each similar to the cross-section of the small column, and spaced 26 1/2 inches apart on centers on the plane of the roof trusses. The separate members of each column are united by solid diagonal sheathing of 2 x 12-inch boards, inclined in different directions on the opposite sides of the column and secured by six sixty-penny spikes in each end of each board.

The main roof trusses are supported by the intermediate rows of columns, a small column at one end of each truss and a large column at the other end. One member of the large column is cut off square to form a seat for the end of the lower chord, and the other member is continued up to form a seat for the top chord. At the other end of the truss the small column is continuous to the seat for the top chord, and a special seat is keyed and bolted to it on the inside to receive the bearing for the lower chord.

The trusses are made with 10-foot panels and a depth of 13 feet on centers. All compression members are of long-leaf southern yellow-pine, and tension members are medium steel round rods with screw ends, most of which are upset. The trusses are seated on framed brackets which project from the intermediate columns and are keyed and bolted to the whole length of the end lower chord panels so as to virtually form long corbels and stiffen the trusses or have some effect of reducing the span.

Excepting the counters in the middle panels, all top-chord and diagonal members are single timbers nearly square in cross-section. At the middle point of the top chord, there is a reinforcement piece keyed and bolted to the under side to receive the diagonal members; at the adjacent top chord panel points and at the middle bottom chord panel points there are cast-iron angle blocks for the diagonals; elsewhere the diagonal members are notched or tenoned into the chords. The bottom chord is lap-jointed in the middle, where it is spliced with two 8 x 5/16-inch steel splice plates 46 inches long with twenty-four 1 1/8-inch bolts. All the nuts on the lower ends of the vertical rods take bearing on standard cast-iron washers. The rods pass through holes bored through the chords and their upper ends are received in holes countersunk in the under sides of the purlins. The upper nuts take bearing on 1/2-inch steel plates which project on both sides of the chords and form seats and splices for the purlins, to which they are spiked. At the ends of the trusses the steel bearing plates are bent to receive the sway-braced diagonal rods. The purlins support 2 x 6-inch rafters, 18 inches apart on centers, which are sheathed with 1-inch boards covered with metal tile and rubberoid. The trusses are cambered 3 inches in the center, and, where reinforcement pieces are keyed to the chord, they have 2-inch oak pins driven tight in holes bored half in each piece and provided with a 3/4-inch through bolt at each key. In the middle of the roof there is a horizontal skylight 40 feet wide covered with a translucent material.

Where the roof trusses have very narrow seats on one side of the 10 x 10-inch column their yellow pine lower chords are reinforced with an oak angle block deeply notched to receive the square end of the inclined end post. A corresponding oak shoe is put opposite to it on the under side of the lower chord and supports it for a length of about 4 feet, thus reducing the unit pressure on the pine and giving a solid bearing on the end of the column member. Both shoe and angle block are keyed to the lower chord, and are secured to it with the same straps and through bolts, which are arranged so as to avoid unnecessary cutting of the cross-section.

Proportioning Footings

It is very important that the foundations, whether continuous, as in a foundation wall, or isolated, as when divided into piers, should have the footing courses proportioned to the weight they will be required to carry, and to the bearing capacity of the soil.

The pressure on the soil from each square foot of the footings should be the same, where the soil is uniform, and at no place must the bearing power of the soil be exceeded. To secure the most satisfactory results, therefore, the footings must be proportioned to properly distribute the weight they are to carry over sufficient areas of ground, to secure uniform settlement in each case. If these conditions were always properly considered, there would be few cracks in the mason work, as such cracks are caused usually by unequal settlement. A uniform settlement even of an inch or more would in most buildings pass unnoticed.

The pressure under a brick pier which supports a tier of columns may be assumed at 10 per cent. less than the calculations show, when the exterior support of the building is a brick wall; for the joints in the brickwork will close slightly under the weight and cause about 10 per cent more settlement than will exist in the columns, each of which, being one piece, is practically jointless, and hence will settle less.

One of the objects in proportioning the footings is to provide for uniform settlement in all parts of the building, so that the floors may remain level, and that no cracks may occur in the walls. Therefore, the loads for which the footings are proportioned should be as near the actual conditions as possible, or, as stated in the Chicago building law, "Foundations shall be proportioned to the actual average loads they will have to carry in the completed and occupied building, and not to the theoretical or occasional loads."

Thus it will be seen that the dead load under the walls of a five-story building would be a considerable item, while the dead load under a tier of iron columns would be much less in proportion to the floor area supported; and as the dead load is always constant, and the live load may vary greatly, only the amount of live load that will probably be supported by the footings should be considered.

For warehouses, stores, etc., 50 percent of the live load that the floorbeams have to carry should be added to the dead load carried on the footings. For office buildings, hotels, dwelling houses, etc., the weight of the people occupying them need not enter into the calculations for proportions of footings, and only from 25 to 80 pounds per square foot of floor need be allowed for the weight of furniture, books, safes, etc. It has been proved by statistics that the average permanent loads do not exceed the above limits. For theaters, halls, etc., a larger allowance should be made for the weight of people, but even a densely packed crowd of men will not weigh more than 100 pounds per square foot of floor.

Arched Systems

Girders and Brick Arches

Among the earliest forms of fireproof floors were those consisting of brick arches of small or moderate span, supported by cast- or wrought-iron girders.

Such constructions are still in use for mills, warehouses, sugar factories, and other buildings, where great weights have to be stored.

The girders may be placed from 4 to 12 feet apart; the arches turned from one to the other; the spandrils filled in and levelled up with concrete, and covered with a floor of any material.

A tension rod, unites the girders, and prevents their yielding under the thrust of the arches. The nearer the tension rod is to the springing of the arch the better, but it is frequently kept high up within the arch in order that it may not be visible.

The tension rod is often used only for the outer arches of a series; these, being thus prevented from yielding, form an abutment for the others. This increase of thickness is useless unless the extra rings are bonded in with the others, or built with bricks the full depth of the arch, which amounts to the same thing.

Sir W. Fairbairn recommends that the rise of such arches should be 1/10 the span for floors of mills, and 1/8 the span for warehouse floors to carry heavy weights.

Whichcord's Fireproof Blocks.1 - It will be seen that in the systems above described the lower surfaces of the iron girders are exposed to the direct action of fire.

To prevent this, the late Mr. Whichcord imbedded them in fireclay blocks, which protect them from fire, and, at the same time, form skewbacks for the arches.

Where ceiling joists are used, they may be supported on the lower ledges of the fire blocks.

In this the blocks or voussoirs of the arch are of hollow fireclay blocks, which are stated to be 1/3 lighter than bricks or concrete. The under sides of these are dovetail grooved so as to form a key for the plastering.

Used at the National Safe Deposit Co.'s Warehouse.

This flooring is capable of sustaining great weights; the girders are well protected; the arches being light may be of considerable span, though large spans increase the depth and cost of the floor, and, moreover, they do exert a certain amount of thrust upon the walls.

This flooring has been used at Whiteleys, The London Pavilion, National Provincial Bank, and in several warehouses.

Northcroft's System 1 consists of flat arches of specially moulded firebricks, resting upon fire-brick skewbacks supported by I iron girders.

These girders rest at the ends upon turned rollers, and are thus permitted to expand when heated.

Dennett's Fireproof Floor consists of concrete arches supported where they abut upon the walls by projecting courses, and at intermediate points by rolled or riveted iron girders.

The arches should have a minimum rise of 1 inch to every foot of width up to spans of 10 or 12 feet, and are sustained by centering until they are thoroughly set.

The concrete used has sulphate of lime (gypsum) for its matrix. It has been proved that this substance does not lose its cohesive power even when it is raised to a white heat and then drenched with cold water.

The floor above the arch may be formed by simply bringing the concrete itself to a smooth surface. Joists may be nailed to fillets laid upon the concrete, in a similar manner. The soffit of the arch may be finished at once with the setting coat of plastering; 1 or, if a flat ceiling is necessary, joists must be fixed to the lower flanges of the girders to carry the lath and plaster.

"Wilkinson's System is very like Dennett's in form, but the arches are of concrete granite and the ceilings formed with fibrous plaster slabs.

This system has been used at Edinburgh University, several stations on North-Eastern Railway, by the War Department, and in many warehouses. Lindsay's Arch System consists of "Pumice concrete" arches resting upon girders which are stiffened by truss rods, and whose lower flange is protected from fire by an iron trough filled in with pumice concrete.

Concrete arches are often laid upon a soffit of corrugated iron, which supports the concrete while it is being laid and protects it afterwards.

Wood and Concrete Floor

Bunnett's System consists of hollow bricks of a peculiar shape laid in a flat arch from wall to wall, resting on angle irons held together by a tension rod passing through the bricks. The bricks are so arranged and formed laterally that each receives the support of six adjoining bricks. The under sides of the bricks are dovetail grooved to afford a key for the plastering.

Tile Floors consist of arches formed with flat tiles resting upon girders, in the same way as the brick arches above described. When the centering is fixed, the first course of tiles is laid dry, and then covered with cement; upon this a second, third, and fourth course are laid in the same manner. The spandrils are filled in with concrete, and the floor finished with joists and boarding, pavement, or in any way that may be desired.

Disadvantages of Arched Floors

Arched floors, especially when composed of voussoir blocks, are complicated and heavy. The arches depend greatly upon one another, and if one gives way it may lead to the failure of the whole floor. In any case they exert a thrust upon the walls.

These disadvantages have led to the adoption of simpler forms of fireproof flooring.

First and Second Class Buildings

Every building hereafter erected more than seventy-five feet in height, or hereafter increased in height to more than seventy-five feet, shall be a first class building. Every second class building hereafter erected more than four stories in height, and any second class building now in existence and increased in height to more than four stories shall have the first floor and the basement and cellar stories of first class construction except as hereinafter otherwise mentioned for buildings for habitation. Every building for habitation hereafter erected covering more than five thousand square feet, or more than five stories in height, shall be a first class building. Every building altered or enlarged and occupied or to be occupied as a habitation, to be in excess of sixty-five feet in height, or in excess of five thousand square feet in superficial area, or in excess of five stories in height, shall be a first class building. Every building hereafter erected within the building limits to be occupied as a permanent schoolhouse shall be a first class building. Every building hereafter erected as a theatre, and every building hereafter altered to be occupied as a theatre, shall be a first class building. Every building hereafter erected for, altered to or converted to use as a moving picture house shall be a first class building. All other buildings may be of second or third class construction.

Except as is otherwise provided herein, buildings adapted for habitations, and not more than five stories in height, may be erected, remodelled or enlarged of second class construction, but no such building shall exceed five thousand square feet in superficial area or sixty-five feet in height. Every such building exceeding thirty-five hundred square feet of superficial area, and every such building exceeding seventeen hundred and fifty square feet in superficial area, and more than four stories or fifty-five feet in height, shall have the first floor and basement and cellar stories of first class construction, with no openings through the first floor, except for piping; provided that stairways from the first story to the outside may penetrate the floor construction. Said stairs herein referred to are to be fireproof, separated from the basement or cellar by walls of solid masonry at least eight inches thick, with no opening to cellar or basement. The first story, or basement, or both the first story and basement, in such building more than seventeen hundred and fifty square feet in superficial area so constructed, remodelled or enlarged, may be used for mercantile purposes; provided that the floors and walls separating the portion of the building used for mercantile purposes and the portion used for habitation be of first class construction with no openings, except for piping, but that stairways from the portion used for habitation to the outside may penetrate the mercantile portion. In such event, the stairs herein referred to are to be constructed of incombustible material and separated from the latter by solid masonry walls at least eight inches thick with no opening to the portion used for habitation. In such buildings where the second floor is of first class construction the requirement providing for the first floor, basement and cellar stories to be of first class construction may be omitted.

Restriction of Areas

Any first class building used above the first floor as a warehouse or store for the storage or sale of merchandise shall have all vertical openings protected by fireproof enclosures. Such enclosures shall, if enclosing stairs or escalators, have automatic doors, and all glass in the enclosure shall be wire glass.

Such a building shall be so divided by brick walls built like party walls with the same openings allowed, that no space inside the buildings shall exceed in area ten thousand square feet, except that when any such building has a frontage of not less than fifty feet on each of two streets, such space may exceed ten thousand square feet in area, provided that buildings in which such extension of area beyond ten thousand square feet is permitted shall have automatic fire sprinklers installed and means of ingress and egress satisfactory to the commissioner and the board of appeal.

Second class buildings used above the first floor as warehouses or stores for the storage or sale of merchandise shall be so divided by brick walls, built like party walls with the same openings allowed, that no space inside such buildings shall exceed in area ten thousand square feet, and no existing wall in any second class building shall be removed so as to leave an area of more than ten thousand square feet, nor shall any existing wall separating areas which combined would exceed ten thousand square feet in area, have openings cut in it greater in area or number than is allowed in this act for party walls.

Every second class building more than three stories high and used above the first floor as a warehouse or store for the storage or sale of merchandise, shall have all vertical openings for elevators and stairways, air or light shafts, through its floors protected by fireproof enclosures. Such enclosures shall be supported by fireproof supports and framing, and shall, if enclosing stairs or escalators, have automatic doors, and all glass in the enclosure shall be wire glass.

No building used above the first floor for the storage or sale of merchandise shall have less than two means of egress from every story, one of which means may be either an outside fire escape or through a brick wall closed by automatic doors into a building of the same class; except that an independent monumental stairway extending from the basement to the second floor may be constructed.

Buildings for Manufacturing Purposes

Buildings outside the building limits and adapted exclusively for manufacturing, storage, mechanical or stable purposes, may be built under such conditions as the commissioner shall prescribe. If of wood such buildings shall not exceed forty-five feet in height.

Iron and Steel Supports

For large spans and heavy weights it will be often necessary to use iron or steel columns with wooden girders. If cast-iron columns are used, they should not be made with a shell less than three-quarters of an inch thick. This is necessary on account of the danger of an unequal thickness in the shell of the column. To obtain the hollow column, the casting must be made about a "core," and although this core is accurately centered, there is danger of its being displaced by the pouring in of the molten iron, as, being of a lighter composition, it will have a tendency to "float." This may result in an added thickness to one side of the shell of the column and -a corresponding lack of metal on the opposite side. To guard against this defect, cast-iron columns should always be tested by boring a small hole on opposite sides, and if more than one-fourth of the thickness of shell is wanting in any column it should be rejected. The outer surface of cast iron should be smooth and clean, with sharp angles; and all projections, such as lugs, caps, or bases, should be closely examined to detect the presence of cracks which may occur at these points.

Character of Buildings

The buildings under the control of the Treasury Department are for the use of the civil branches of the Government, such as Postoffices, Courts, Customs, Internal Revenue, Marine Hospitals, and Quarantine Stations. Such buildings as the Capitol, the Library of Congress, and the Executive buildings of the different Departments at Washington, are not under the control of the Treasury Department nor built by it; neither are forts or other Army or Navy buildings, or United States Jails.

Many of the buildings under the Treasury Department are very large and expensive, the New York Custom House and the Chicago and San Francisco Postoffices being buildings of this type. The larger number, however, are of moderate size, ranging from 40 feet by 80 feet outside dimensions, and one story high, to buildings of three times that area and three stories high, and costing complete, with heating apparatus, etc., from $35,000 to $175,000 each. The large majority are of fireproof construction and classic design, with interior finish of hardwood and marble where the funds will permit. The best of materials and appliances are used throughout, as it is inadvisable and almost impracticable in public buildings of this character, to exercise the many economies which prevail in private building.

The interiors are fitted with the dest of modern appliances and conveniences; plumbing is extensive and perfect; the toilet rooms for Postoffice carriers are supplied not only with the usual appliances, but often with shower-baths; and no woodwork either in floors or in finish is allowed. Through the building, small private toilet rooms are provided for the more prominent officials; and in many, especially in the South, bathtubs are supplied. The heating, ventilating, and electric work are, in their lines, equally complete.

The buildings in the larger cities where the Railway Mail Service men change, have dormitories for the accommodation of from ten to seventy-five men. In the Postoffice proper are galleries with openings commanding every corner accessible to the men, from which Inspectors can watch for days, without arousing suspicion, any person suspected of pillaging the mails.

There is always a large "swing room" for carriers when off duty, where they provide themselves with games, boxing gloves, and occasionally a billiard table for recreation.

In the larger Postoffice buildings are the most perfect of mechanical equipments, engines, boilers, electric generators, elevators, mail-carrying and handling devices, sweeping machinery, etc.

From the above it is evident that there are few lines in the building industries which should not be interested in the erection and furnishing of these many buildings.

Roof Characteristics

Styles of Roofs

The different varieties of roofs, from the simple pitch roof to the most complicated combination of hips and valleys, are developments of a few simple forms.

Lean-to Roof

The lean-to roof is the most simple of them all, and is usually employed for small sheds, piazzas, porches, ells, and in many other situations where appearance is not a matter of great moment, and where the essential thing is to obtain the shelter as cheaply and as easily as possible. The position in which the surface is supported enables the rain water to drain freely from it, and thus it fulfills the requirements of a roof so long as it remains water-tight.

6

Pipe Flow

Pipe flow, a branch of Hydraulics and Fluid Mechanics, is a type of liquid flow within a closed conduit (conduit in the sense of a means of containment). The other type of flow within a conduit is open channel flow. These two types of flow are similar in many ways, but differ in one important respect. Pipe flow does not have a free surface which is found in open-channel flow. Pipe flow, being confined within closed conduit, does not exert direct atmospheric pressure, but does exert hydraulic pressure on the conduit.

Not all flow within a closed conduit is considered pipe flow. Storm sewers are closed conduits but usually maintain a free surface and therefore are considered open-channel flow. The exception to this is when a storm sewer operates at full capacity, and then can become pipe flow. Energy in pipe flow is expressed as head and is defined by the Bernoulli equation. In order to conceptualize head along the course of flow within a pipe, diagrams often contain a hydraulic grade line. Pipe flow is subject to frictional losses as defined by the Darcy-Weisbach formula.

States of Flow

The behaviour of pipe flow is governed mainly by the effects of viscosity and gravity relative to the inertial forces of the flow. Depending on the effect of viscosity relative to inertia, as represented by the Reynolds number, the flow can be either laminar or turbulent. At a Reynolds number below the critical value of approximately 2040 pipe flow will ultimately be laminar, whereas above the critical value turbulent flow can persist. In addition, the transition between laminar flow and turbulence can be sensitive to disturbance levels and imperfections.

Flow through pipes can roughly be divided into two:

- Laminar flow
- Turbulent flow

Hagen–Poiseuille Equation

In fluid dynamics, the Hagen–Poiseuille equation, also knwn as the Hagen–Poiseuille law, Poiseuille law or Poiseuille equation, is a physical law hat gives the pressure drop in a fluid flowing through a long cylindrical pipe. It can be successfully applied to air flow in lung alveoli, for the flow through a drinking straw or through a hypodermic needle. It was experimentally derived independently by Gotthilf Heinrich Ludwig Hagen in 839 and Jean Louis Marie Poiseuille in 1838, and published by Poiseuille in 1840 and 1846.

The assumptions o the equation are that the fluid is incompressible and newtonian; the flow s laminar through a pipe of constant circular cross-section that is substantially longer than its diameter; and there is no acceleration of fluid in the pipe. For velocities and pipe diameters above a threshold, actual fluid flow is not laminar but turbulent, leading to larger pressure drops than calculated by the Hagen–Poiseuille equation.

Euation

Standard fluid dynamics notation:

In standard fluid dynamics notation:

$$\Delta P = \frac{8\mu L Q}{\pi r^4}$$

or

$$\Delta P = \frac{128\mu L Q}{\pi d^4}$$

where:

ΔP is the pressure loss

$L$is the length of pipe

μ is the dynamic viscosity

Q is the volumetric flow rate

r is the radius

d is the diameter

π is the mathematical constant Pi

Physics Notation

$$\Phi = \frac{dV}{dt} = v\pi R^2 = \frac{\pi R^4}{8\eta}\left(\frac{-\Delta P}{\Delta x}\right) = \frac{\pi R^4}{8\eta}\frac{|\Delta P|}{L}$$

where in compatible units (e.g., S.I.):

Φ is the volumetric flow rate (denoted as Q above)

$V(t)$ is the volume of the liquid transferred as a function of time,

v is mean fluid velocity along the length of the tube

x is distance in direction of flow

R is the internal radius of the tube

ΔP is the pressure difference between the two ends

is the dynamic fluid viscosity (S.I. unit: pascal-second (Pa·s)),

L is the length of the tube

The equation does not hold close to the pipe entrance.

The equation fails in the limit of low viscosity, wide and/or short pipe. Low viscosity or a wide pipe may result in turbulent flow, making it necessary to use more complex models, such as Darcy-Weisbach equation. If the pipe is too short, the Hagen–Poiseuille equation may result in uphysically high flow rates; the flow is bounded by Bernoulli's principle, under less restrictive conditions, by

$$\Phi_{max} = \pi R^2 \sqrt{2\Delta P / \rho}\,.$$

Relation to Darcy–Weisbach

This result is also a solution to the phenomenological Darcy–Weisbach equation in the field of hydraulics, given a relationship for the friction factor in terms of the Reynolds number:

$$\Lambda = \frac{64}{\mathrm{Re}}, \qquad \mathrm{Re} = \frac{2\rho v r}{\eta},$$

where *Re* is the Reynolds number and ρ fluid density. In this form the law approximates the *Darcy friction factor*, the *energy (head) loss factor*, *friction loss factor* or *Darcy (friction) factor* Ë in the laminar flow at very low velocities in cylindrical tube. The theoretical derivation of a slightly different form of the law was made independently by Wiedman in 1856 and Neumann and E. Hagenbach in 1858 (1859, 1860). Hagenbach was the first who called this law the Poiseuille's law.

The law is also very important specially in hemorheology and hemodynamics, both fields of physiology.

The Poiseuilles' law was later in 1891 extended to turbulent flow by L. R. Wilberforce, based on Hagenbach's work.

Liquid Flow through a Pipe

Assume the liquid exhibits Laminar flow. Laminar flow in a round pipe prescribes that there are a bunch of circular layers (lamina) of liquid, each having a velocity determined only by their radial distance from the centre of the tube. Also assume the centre is moving fastest while the liquid touching the walls of the tube is stationary (due to the no-slip condition).

To figure out the motion of the liquid, all forces acting on each lamina must be known:

1. The pressure force pushing the liquid through the tube is the change in pressure multiplied by the area: $F = -\Delta PA$. This force is in the direction of the motion of the liquid. The negative sign comes from the conventional way we define $\Delta P = P_{end} - P_{top} < 0$.
2. Viscosity effects will pull from the faster lamina immediately closer to the centre of the tube.
3. Viscosity effects will drag from the slower lamina immediately closer to the walls of the tube.

Viscosity

Figure: *Two fluids moving past each other in the* x *direction. The liquid on top is moving faster and will be pulled in the negative direction by the bottom liquid while the bottom liquid will be pulled in the positive direction by the top liquid.*

When two layers of liquid in contact with each other move at different speeds, there will be a shear force between them. This force is proportional to the area of contact A, the velocity gradient in the direction of flow $\Delta v_x / \Delta y$, and a proportionality constant η (viscosity) and is given by

$$F_{\text{viscosity, top}} = -\eta A \frac{\Delta v_x}{\Delta y}.$$

The negative sign is in there because we are concerned with the faster moving liquid (top in figure), which is being slowed by the slower liquid (bottom in figure). By Newton's third law of motion, the force on the slower liquid is equal and opposite (no negative sign) to the force on the faster liquid. This equation assumes that the area of contact is so large that we can ignore any effects from the edges and that the fluids behave as Newtonian fluids.

Faster Lamina

Assume that we are figuring out the force on the lamina with radius *s*. From the equation above, we need to know the area of contact and the velocity gradient. Think of the lamina as a ring of radius *r*, thickness *dr*, and length Δx. The area of contact between the lamina and the faster one is simply the area of the inside of the cylinder: $A = 2\pi r \Delta x$. We don't know the exact form for the velocity of the liquid within the tube yet, but we do know (from our assumption above) that it is dependent on the radius. Therefore, the velocity gradient is the change of the velocity with respect to the change in the radius at the intersection of these two laminae. That intersection is at a radius of *r*. So, considering that this force will be positive with respect to the movement of the liquid (but the derivative of the velocity is negative), the final form of the equation becomes

$$F_{\text{viscosity, fast}} = -\eta 2\pi r \Delta x \left.\frac{dv}{dr}\right|_r$$

where the vertical bar and subscript *r* following the derivative indicates that it should be taken at a radius of *r*.

Slower Lamina

Next let's find the force of drag from the slower lamina. We need to calculate the same values that we did for the force from the faster lamina. In this case, the area of contact is at *r+dr* instead of *r*. Also, we need to remember that this force opposs the direction of movement of the liquid and will therefore be negative (and that the derivative of the velocity is negative).

$$F_{\text{viscosity, slow}} = \eta 2\pi (r + dr) \Delta x \left.\frac{dv}{dr}\right|_{r+dr}$$

Putting it All Together

To find the solution for the flow of liquid through a tube, we need to make one last assumption. There is no acceleration of liquid in the pipe, and by Newton's first law, there is no net force. If there is no net force then we can add all of the forces together to get zero

$$0 = F_{\text{pressure}} + F_{\text{viscosity, fast}} + F_{\text{viscosity, slow}}$$

or

$$0 = -\Delta P 2\pi r dr - \eta 2\pi r \Delta x \left.\frac{dv}{dr}\right|_{r} + \eta 2\pi (r + dr)\Delta x \left.\frac{dv}{dr}\right|_{r+dr}.$$

First, to get everything happening at the same point, use the first two terms of a Taylor series expansion of the velocity gradient:

$$\left.\frac{dv}{dr}\right|_{r+dr} = \left.\frac{dv}{dr}\right|_{r} + \left.\frac{d^2v}{dr^2}\right|_{r} dr.$$

The expression is valid for all laminae. Grouping like terms and dropping the vertical bar since all derivatives are assumed to be at radius r,

$$0 = -\Delta P 2\pi r dr + \eta 2\pi dr \Delta x \frac{dv}{dr} + \eta 2\pi r dr \Delta x \frac{d^2v}{dr^2} + \eta 2\pi (dr)^2 \Delta x \frac{d^2v}{dr^2}.$$

Finally, put this expression in the form of a differential equation, dropping the term quadratic in dr.

$$\frac{1}{\eta}\frac{\Delta P}{\Delta x} = \frac{d^2v}{dr^2} + \frac{1}{r}\frac{dv}{dr}$$

It can be seen that both sides of the equations are negative: there is a drop of pressure along the tube (left side) and both first and second derivatives of the velocity are negative (velocity has a maximum value at the centre of the tube, where $r = 0$). Using the chain rule, the equation may be re-arranged to:

$$\frac{1}{\eta}\frac{\Delta P}{\Delta x} = \frac{1}{r}\frac{d}{dr} r \frac{dv}{dr}.$$

This differential equation is subject to the following boundary conditions:

$v(r) = 0$ at $r = R$— "no-slip" boundary condition at the wall

$\frac{dv}{dr} = 0$ at $r = 0$— axial symmetry.

Axial symmetry means that the velocity v(r) is maximum at the centre of the tube, therefore the first derivative $\frac{dv}{dr}$ is zero at r = 0.

The differential equation can be integrated to:

$$v(r) = \frac{1}{4\eta} r^2 \frac{\Delta P}{\Delta x} + A\ln(r) + B.$$

To find A and B, we use the boundary conditions.

First, the symmetry boundary condition indicates:

$$\frac{dv}{dr} = \frac{1}{2\eta} r \frac{\Delta P}{\Delta x} + A\frac{1}{r} = 0 \text{ at r = 0.}$$

A solution possible only if A = 0. Next the no-slip boundary condition is applied to the remaining equation:

$$v(R) = \frac{1}{4\eta} R^2 \frac{\Delta P}{\Delta x} + B = 0$$

so therefore

$$B = -\frac{1}{4\eta} R^2 \frac{\Delta P}{\Delta x}.$$

Now we have a formula for the velocity of liquid moving through the tube as a function of the distance from the centre of the tube

$$v = -\frac{1}{4\eta}\frac{\Delta P}{\Delta x}(R^2 - r^2)$$

or, at the centre of the tube whre the liquid is moving fastest (r = 0) with R being the radius of the tube,

$$v_{max} = -\frac{1}{4\eta}\frac{\Delta P}{\Delta x} R^2.$$

Poiseuille's Law

To get the total volume that flows through the tube, we need to add up the contributions from each lamina. To calculate the flow through each lamina, we multiply the velocity (from above) and the area of the lamina.

$$\Phi(r)dr = \frac{1}{4\eta}\frac{|\Delta P|}{\Delta x}(R^2 - r^2)2\pi r dr = \frac{\pi}{2\eta}\frac{|\Delta P|}{\Delta x}(rR^2 - r^3)dr$$

Finally, we integrate over all lamina via the radius variable r.

$$\Phi = \frac{\pi}{2\eta}\frac{|\Delta P|}{\Delta x}\int_0^R (rR^2 - r^3)dr = \frac{|\Delta P|\pi R^4}{8\eta\Delta x}$$

Poiseuille's Equation for Compressible Fluids

For a compressible fluid in a tube the volumetric flow rate and the linear velocity is not constant along the tube. The flow is usually expressed at outlet pressure. As fluid is compressed or expands, work is done and the fluid is heated and cooled. This means that the flow rate depends on the heat transfer to and from the fluid. For an ideal gas in the isothermal case, where the temperature of the fluid is permitted to equilibrate with its surroundings, and when the pressure difference between ends of the pipe is small, the volumetric flow rate a the pipe outlet is given by

$$\Phi = \frac{dV}{dt} = v\pi R^2 = \frac{\pi R^4 \left(P_i - P_o\right)}{8\eta L} \times \frac{P_i + P_o}{2P_o} = \frac{\pi R^4}{16\eta L}\left(\frac{P_i^2 - P_o^2}{P_o}\right)$$

where

P_i inlet pressure

P_o outlet pressure

L is the length of tube

η is the viscosity

R is the radius

V is the volume of the fluid at outlet pressure

v is the velocity of the fluid at outlet pressure

This is usually a good approximation when the flow velocity is less than mach 0.3

This equation can be seen as Poiseuille's law with an extra correction factor $\frac{P_i + P_o}{2} \times \frac{1}{P_o}$ expressing the average pressure relative to the outlet pressure.

Electrical Circuits Analogy

Electricity was originally understood to be a kind of fluid. This hydraulic analogy is still conceptually useful for understanding circuits. This analogy is also used to study the frequency response of fluid mechanical networks using circuit tools, in which case the fluid network is termed a hydraulic circuit.

Poiseuille's law corresponds to Ohm's law for electrical circuits ($V = IR$), where the pressure drop ΔP is analogous to the voltage V and volumetric flow rate Φ is analogous to the current I. Then the resistance

$$R = \frac{8\eta\Delta x}{\pi r^4}.$$

This concept is useful because the effective resistance in a tube is inversely proportional to the fourth power of the radius. This means that halving the radius of the tube increases the resistance to fluid movement by a factor of 16. Both Ohm's law and Poiseuille's law illustrate transport phenomena.

Moody Chart

The Moody chart or Moody diagram is a graph in non-dimensional form that relates the Darcy-Weisbach friction factor, Reynolds number and relative roughness for fully developed flow in a circular pipe. It can be used for working out pressure drop or flow rate down such a pipe.

Description

This dimensionless chart is used to work out pressure drop, ΔP (Pa) (or head loss, h_f (m)) and flow rate through pipes. Head loss can be calculated using the Darcy–Weisbach equation:

$$h_f = f\frac{l}{d}\frac{V^2}{2g};$$

(not to be confused with the Fanning equation and the Fanning friction factor:

$$h_f = 4f\frac{l}{d}\frac{V^2}{2g},$$

which uses a friction-factor equal to one fourth the Darcy-Weisbach friction factor).

Pressure drop can then be evaluated as:

$$\Delta P = \rho g h_f \text{ or directly from } \Delta P = f\frac{\rho V^2}{2}\frac{l}{d},$$

where ρ is the density of the fluid, V is the average velocity in the pipe, f is the friction factor from the Moody chart, l is the length of the pipe and d is the pipe diameter.

The basic chart plots Darcy–Weisbach friction factor against Reynolds number for a variety of relative roughnesses and flow regimes. The relative roughness being the ratio of the mean height of roughness of the pipe to the pipe diameter or $\frac{\epsilon}{d}$.

The Moody chart can be divided into two regimes of flow: laminar and turbulent. For the laminar flow regime, the Darcy–Weisbach friction factor was determined analytically by Poiseuille and $\frac{64}{Re}$ is used. In this regime roughness has no discernible effect. For the turbulent flow regime, the relationship between the friction factor and the Reynolds number is more complex and is governed by the Colebrook equation which is implicit in f:

$$\frac{1}{\sqrt{f}} = -2.0\log_{10}\left(\frac{\frac{\epsilon}{d}}{3.7} + \frac{2.51}{Re\sqrt{f}}\right), \text{turbulent flow.}$$

In 1944, Lewis Ferry Moody plotted the Darcy–Weisbach friction factor into what is now known as the Moody chart. The Fanning friction factor is 1/4 the Darcy–Weisbach one and the equation for pressure drop has a compensating factor of four.

Open-Channel Flow

Open-channel flow, a branch of hydraulics, is a type of liquid flow within a conduit with a free surface, known as a channel. The other type of flow within a conduit is pipe flow. These two types of flow are similar in many ways, but differ in one important respect: the free surface. Open-channel flow has a free surface, whereas pipe flow does not.

Classifications of Flow

Open-channel flow can be classified and described in various ways based on the change in flow depth with respect to time and space. The fundamental type of flow dealt with in open-channel hydraulics.

Time as the Criterion

- *Steady flow* – The depth of flow does not change over time, or if it can be assumed to be constant during the time interval under consideration.
- *Unsteady flow* - The depth of flow does change with time.

Space as the Criterion

- *Uniform flow*: The depth of flow is the same at every section of the channel. Uniform flow can be steady or unsteady, depending on whether or not the depth changes with time, (although unsteady uniform flow is rare).

- *Varied flow:* The depth of flow changes along the length of the channel. Varied flow technically may be either steady or unsteady.
- Varied flow can be further classified as either rapidly or gradually varied.

Rapidly varied flow: The depth changes abruptly over a comparatively short distance. Rapidly varied flow is known as local phenomenon. Examples are the hydraulic jump and the hydraulic drop.

Gradually varied flow: The depth changes over a long distance.

Continuous flow: The discharge is constant throughout the reach of the channel under consideration. This is often the case with steady flow. This flow is considered continuous and therefore can be described using the continuity equation for continuous steady flow.

Spatially varied or discontinuous flow - The discharge of a steady flow is nonuniform along a channel. This happens when water enters and/or leaves the channel along the course of flow. An example of flow entering a channel would be a road side gutter. An example of flow leaving a channel would be an irrigation channel. This flow can be described using the continuity equation for continuous unsteady flow requires the consideration of the time effect and includes a time element as a variable.

States of Flow

The behaviour of open channel flow is governed by the effects of viscosity and gravity relative to the inertial forces of the flow. Surface tension has a minor contribution, but does not play a significant enough role in most circumstances to be a governing factor. Depending on the effect of viscosity relative to inertia, as represented by the Reynolds number, the flow can be either laminar, turbulent, or transitional.

Surge Control

Surge control is used to prevent pressure surges, which can occur in a hydraulic system when the flow in a pipeline is stopped too quickly. When a valve closes or a pump stops the water or oil or other fluid in a hydraulic system will attempt to continue flowing. This will cause an increase in pressure which can damage the pipe. Surge control devices absorb and limit this pressure increase and prevent damage to the pipework.

What is Hydraulic Surge

Hydraulic pressure (transient flow) occurs when the flow of fluid in a pipeline is abruptly changed. Piping systems that use quick acting valves, or use pumps that start up or shut down rapidly, are susceptible to pressure transients (surges). These conditions can result in piping failure, damage to pumps, fittings, instrumentation, and other system components.

Controlling Hydraulic Surge

Controlling hydraulic surges can increase the life of hydraulic systems by effectively managing these destructive forces. Fast acting surge relief valves, Surge suppressors, bladder surge tanks, and hydropneumatic device (surge tank) can be used to control surges. Please refer to article "Specifying surge relief valves in liquid pipelines" with Intec Journal July/August 2012 of ISA.

Typical Applications

Surge control products are currently protecting fluid systems in numerous industries and defence industries, including airport refuelling systems, bulk marine loading/unloading facilities, hydrocarbon pipeline system, fire systems (as surge tanks), municipal water and sewage pump systems, petro-chemical facilities, cogeneration plants, mobile fuel trucks, and diesel fuel loading

Surge Relief Valves

Surge Relief Valves are used in most applications including marine loading/unloading application, liquid pipeline transfer systems, fire protection and water pumping stations. They utilise a differential pressure valve to control pressure transients. A surge relief valve for a liquid transporting pipeline, comprising a surge relief line adapted to be tapped from the main transportation pipeline. The relief line includes the surge relief valve which is conditioned to open very quickly to substantial flow when the pressure in the line exceeds a predetermined safe limit, which is well above the normal pipeline pressure.

When the surge relief valve opens very quickly, it diverts to a reservoir called the surge relief tank (Floating roof, slop tank, closed tank etc.), that quantity of liquid which is necessary to prevent the pressure from exceeding said predetermined limit. Hence, in the event of the surge, much of the pressure is absorbed in the liquid and pipe, and just that quantity of liquid which is necessary to relieve pressures of unsafe proportions is discharged to the surge relief tank. A return line can be employed to evacuate the reservoir to the main pipeline when the surge subsides.

Hydropneumatic Surge Tanks for Water, Waste Water, and Sewage Service

Water, waste water, and sewage pumping applications, incorporate a hydraulically designed nozzle to dampen destructive energy and prevent vortexing.

Hydraulic Machinery

Hydraulic machines are machinery and tools that use liquid fluid power to do simple work. Heavy equipment is a common example.

In this type of machine, hydraulic fluid is transmitted throughout the machine to various hydraulic motors and hydraulic cylinders and which becomes pressurised according to the resistance present. The fluid is controlled directly or automatically by control valves and distributed through hoses and tubes. The popularity of hydraulic machinery is due to the very large amount of power that can be transferred through small tubes and flexible hoses, and the high power density and wide array of actuators that can make use of this power. Hydraulic machinery is operated by the use of hydraulics, where a liquid is the powering medium.

Force and Torque Multiplication

A fundamental feature of hydraulic systems is the ability to apply force or torque multiplication in an easy way, independent of the distance between the input and output, without the need for mechanical gears or levers, either by altering the effective areas in two connected cylinders or the effective displacement (cc/rev) between a pump and motor. In normal cases, hydraulic ratios are combined with a mechanical force or torque ratio for optimum machine designs such as boom movements and trackdrives for an excavator.

Examples

Two Hydraulic Cylinders Interconnected: Cylinder C1 is one inch in radius, and cylinder C2 is ten inches in radius. If the force exerted on C1 is 10 lbf, the force exerted by C2 is 1000 lbf because C2 is a hundred times larger in area ($S = \pi r^2$) as C1. The downside to this is that you have to move C1 a hundred inches to move C2 one inch. The most common use for this is the classical hydraulic jack where a pumping cylinder with a small diameter is connected to the lifting cylinder with a large diameter.

Pump and Motor

If a hydraulic rotary pump with the displacement 10 cc/rev is connected to a hydraulic rotary motor with 100 cc/rev, the shaft torque required to drive the pump is 10 times less than the torque available at the motor shaft, but the shaft speed (rev/min) for the motor is 10 times less than the pump shaft speed. This combination is actually the same type of force multiplication as the cylinder example (1) just that the linear force in this case is a rotary force, defined as torque.

Both these examples are usually referred to as a hydraulic transmission or hydrostatic transmission involving a certain hydraulic "gear ratio".

Hydraulic Circuits

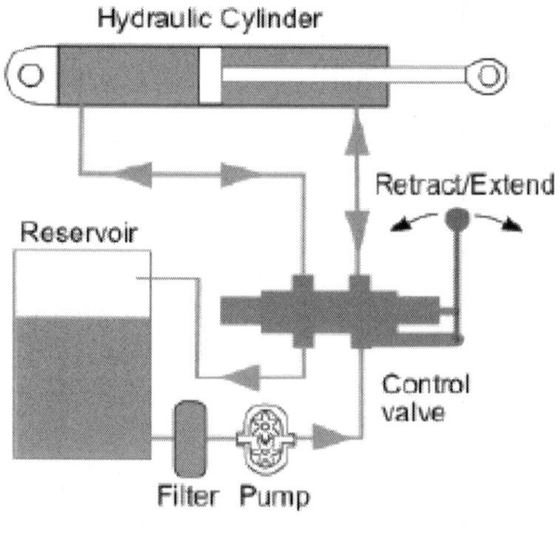

Figure: *A simple* open centre *hydraulic circuit.*

For the hydraulic fluid to do work, it must flow to the actuator and/or motors, then return to a reservoir. The fluid is then filtered and re-pumped. The path taken by hydraulic fluid is called a hydraulic circuit of which there are several types. Open centre circuits use pumps which supply a continuous flow. The flow is returned to *tank* through the control valve's *open centre*; that is, when the control valve is centred, it provides an open return path to tank and the fluid is not pumped to a high pressure. Otherwise, if the control valve is actuated it routes fluid to and from an actuator and tank. The fluid's pressure will rise to meet any resistance, since the pump has a constant output. If the pressure rises too high, fluid returns to tank through a pressure relief valve. Multiple control valves may be stacked in series. This type of circuit can use inexpensive, constant displacement pumps.

Closed centre circuits supply full pressure to the control valves, whether any valves are actuated or not. The pumps vary their flow rate, pumping very little hydraulic fluid until the operator actuates a valve. The valve's spool therefore doesn't need an open centre return path to tank. Multiple valves can be connected in a parallel arrangement and system pressure is equal for all valves.

Constant Pressure and Load-Sensing Systems

The closed centre circuits exist in two basic configurations, normally related to the regulator for the variable pump that supplies the oil:

Constant pressure systems (CP-system), standard. Pump pressure always equals the pressure setting for the pump regulator. This setting must cover the maximum required load pressure. Pump delivers flow according to required sum of flow to the consumers. The CP-system generates large power losses if the machine works with large variations in load pressure and the average system pressure is much lower than the pressure setting for the pump regulator. CP is simple in design. Works like a pneumatic system. New hydraulic functions can easily be added and the system is quick in response.

Constant pressure systems (CP-system), unloaded. Same basic configuration as 'standard' CP-system but the pump is unloaded to a low stand-by pressure when all valves are in neutral position. Not so fast response as standard CP but pump lifetime is prolonged.

Load-sensing systems (LS-system) generates less power losses as the pump can reduce both flow and pressure to match the load requirements, but requires more tuning than the CP-system with respect to system stability. The LS-system also requires additional logical valves and compensator valves in the directional valves, thus it is technically more complex and more expensive than the CP-system. The LS-system system generates a constant power loss related to the regulating pressure drop for the pump regulator:

$$Powerloss = \Delta p_{LS} \cdot Q_{tot}$$

The average Δp_{LS} is around 2 MPa (290 psi). If the pump flow is high the extra loss can be considerable. The power loss also increases if the load pressures vary a lot. The cylinder areas, motor displacements and mechanical torque arms must be designed to match load pressure in order to bring down the power losses. Pump pressure always equals the maximum load pressure when several functions are run simultaneously and the power input to the pump equals the (max. load pressure + Δp_{LS}) x sum of flow.

Five Basic Types of Load-Sensing Systems

- Load sensing *without compensators* in the directional valves. Hydraulically controlled LS-pump.
- Load sensing *with up-stream compensator* for each connected directional valve. Hydraulically controlled LS-pump.
- Load sensing *with down-stream compensator* for each connected directional valve. Hydraulically controlled LS-pump.
- Load sensing *with a combination of up-stream and down-stream compensators*. Hydraulically controlled LS-pump.
- Load sensing with synchronized, both electric controlled pump displacement and electric controlled valve flow area for faster response, increased stability and fewer system losses. This is a new type of LS-system, not yet fully developed.

Technically the down-stream mounted compensator in a valveblock can physically be mounted "up-stream", but work as a down-stream compensator.

System type (3) gives the advantage that activated functions are synchronized independent of pump flow capacity. The flow relation between 2 or more activated functions remains independent of load pressures, even if the pump reaches the maximum swivel angle. This feature is important for machines that often run with the pump at maximum swivel angle and with several activated functions that must be synchronized in speed, such as with excavators. With type (4) system, the functions with *up-stream* compensators have priority. Example: Steering-function for a wheel loader. The system type with down-stream compensators usually have a unique trademark depending on the manufacturer of the valves, for example "LSC" (Linde Hydraulics), "LUDV" (Bosch Rexroth Hydraulics) and "Flowsharing" (Parker Hydraulics) etc. No official standardized name for this type of system has been established but Flowsharing is a common name for it.

Open and Closed Circuits

Open-loop: Pump-inlet and motor-return (via the directional valve) are connected to the hydraulic tank. The term loop applies to feedback; the more correct term is open versus closed "circuit". Open centre circuits use pumps which supply a continuous flow. The flow is returned to tank through the control valve's open centre; that is, when the control valve is centred, it provides an open return path to tank and the fluid is not pumped to a high pressure. Otherwise, if the control valve is actuated it routes fluid to and from an actuator and tank. The fluid's pressure will rise to meet any resistance, since the pump has a constant output. If the pressure rises too high, fluid returns to tank through a pressure relief valve. Multiple control valves may be stacked in series. This type of circuit can use inexpensive, constant displacement pumps.

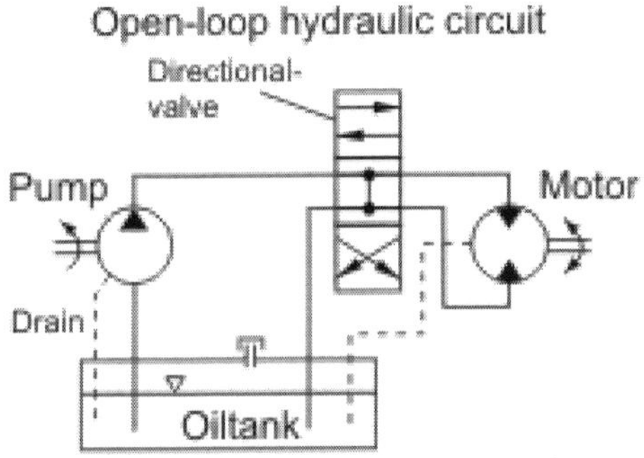

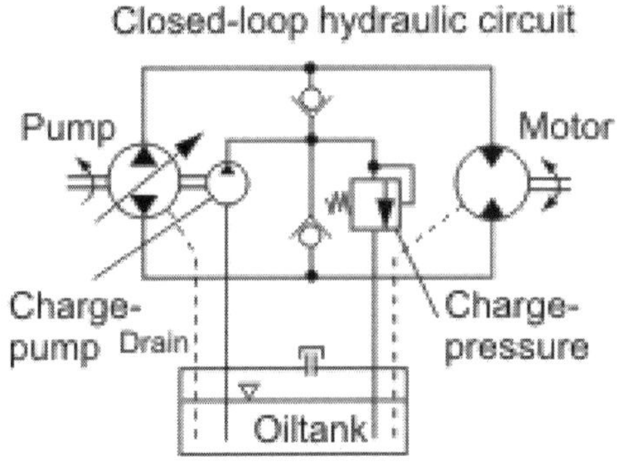

Figure: *Open loop and closed loop circuits*

Closed-loop: Motor-return is connected directly to the pump-inlet. To keep up pressure on the low pressure side, the circuits have a charge pump (a small gearpump) that supplies cooled and filtered oil to the low pressure side. Closed-loop circuits are generally used for hydrostatic transmissions in mobile applications. *Advantages:* No directional valve and better response, the circuit can work with higher pressure. The pump swivel angle covers both positive and negative flow direction. *Disadvantages:* The pump cannot be utilised for any other hydraulic function in an easy way and cooling can be a problem due to limited exchange of oil flow. High power closed loop systems generally must have a 'flush-valve' assembled in the circuit in order to exchange much more flow than the basic leakage flow from the pump and the motor, for increased cooling and filtering. The flush valve is normally integrated in the motor housing to get a cooling effect for the oil that is rotating in the motor housing itself. The losses in the motor housing from rotating effects and losses in the ball bearings can be considerable as motor speeds will reach 4000-5000 rev/min or even more at maximum vehicle speed.

The leakage flow as well as the extra flush flow must be supplied by the charge pump. A large charge pump is thus very important if the transmission is designed for high pressures and high motor speeds. High oil temperature is usually a major problem when using hydrostatic transmissions at high vehicle speeds for longer periods, for instance when transporting the machine from one work place to the other. High oil temperatures for long periods will drastically reduce the lifetime of the transmission. To keep down the oil temperature, the system pressure during transport must be lowered, meaning that the minimum displacement for the motor must be limited to a reasonable value. Circuit pressure during transport around 200-250 bar is recommended.

Closed loop systems in mobile equipment are generally used for the transmission as an alternative to mechanical and hydrodynamic (converter) transmissions. The advantage is a stepless gear ratio (continuously variable speed/torque) and a more flexible control of the gear ratio depending on the load and operating conditions. The hydrostatic transmission is generally limited to around 200 kW maximum power, as the total cost gets too high at higher power compared to a hydrodynamic transmission. Large wheel loaders for instance and heavy machines are therefore usually equipped with converter transmissions. Recent technical achievements for the converter transmissions have improved the efficiency and developments in the software have also improved the characteristics, for example selectable gear shifting programs during operation and more gear steps, giving them characteristics close to the hydrostatic transmission.

Hydrostatic transmissions for earth moving machines, such as for track loaders, are often equipped with a separate 'inch pedal' that is used to temporarily increase the diesel engine rpm while reducing the vehicle speed in order to increase the available hydraulic power output for the working hydraulics at low speeds and increase the tractive effort. The function is similar to stalling a converter gearbox at high engine rpm. The inch function affects the preset characteristics for the 'hydrostatic' gear ratio versus diesel engine rpm.

Components: Hydraulic Pump

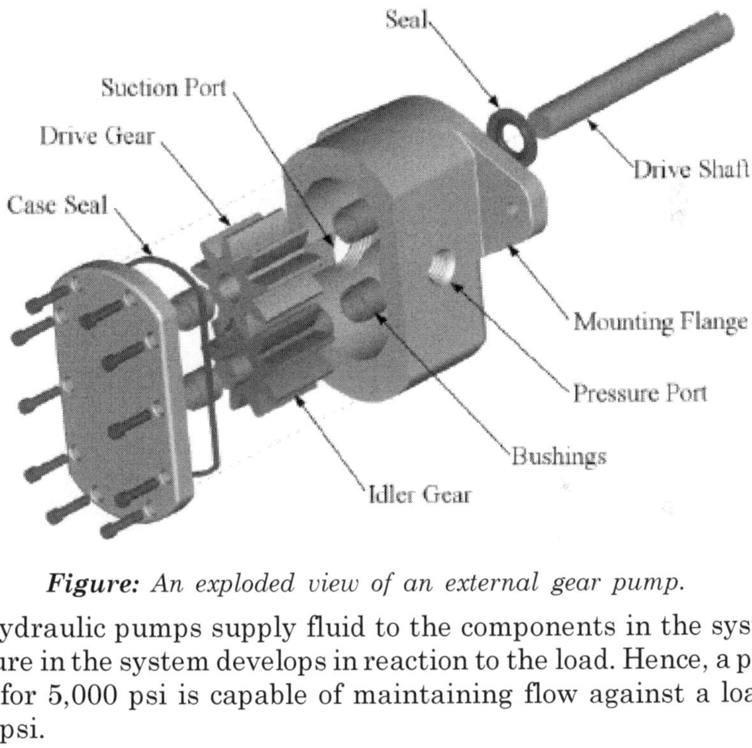

Figure: *An exploded view of an external gear pump.*

Hydraulic pumps supply fluid to the components in the system. Pressure in the system develops in reaction to the load. Hence, a pump rated for 5,000 psi is capable of maintaining flow against a load of 5,000 psi.

Pumps have a power density about ten times greater than an electric motor (by volume). They are powered by an electric motor or an engine, connected through gears, belts, or a flexible elastomeric coupling to reduce vibration.

Common types of hydraulic pumps to hydraulic machinery applications are;

- *Gear pump:* cheap, durable (especially in g-rotor form), simple. Less efficient, because they are constant (fixed) displacement, and mainly suitable for pressures below 20 MPa (3000 psi).

- *Vane pump:* cheap and simple, reliable. Good for higher-flow low-pressure output.
- *Axial piston pump:* many designed with a variable displacement mechanism, to vary output flow for automatic control of pressure. There are various axial piston pump designs, including swashplate (sometimes referred to as a valveplate pump) and checkball (sometimes referred to as a wobble plate pump). The most common is the swashplate pump. A variable-angle swashplate causes the pistons to reciprocate a greater or lesser distance per rotation, allowing output flow rate and pressure to be varied (greater displacement angle causes higher flow rate, lower pressure, and vice versa).
- *Radial piston pump:* normally used for very high pressure at small flows.

Piston pumps are more expensive than gear or vane pumps, but provide longer life operating at higher pressure, with difficult fluids and longer continuous duty cycles. Piston pumps make up one half of a hydrostatic transmission.

Control Valves

Directional control valves route the fluid to the desired actuator. They usually consist of a spool inside a cast iron or steel housing. The spool slides to different positions in the housing, and intersecting grooves and channels route the fluid based on the spool's position.

The spool has a central (neutral) position maintained with springs; in this position the supply fluid is blocked, or returned to tank. Sliding the spool to one side routes the hydraulic fluid to an actuator and provides a return path from the actuator to tank. When the spool is moved to the opposite direction the supply and return paths are switched. When the spool is allowed to return to neutral (centre) position the actuator fluid paths are blocked, locking it in position.

Directional control valves are usually designed to be stackable, with one valve for each hydraulic cylinder, and one fluid input supplying all the valves in the stack.

Tolerances are very tight in order to handle the high pressure and avoid leaking, spools typically have a clearance with the housing of less than a thousandth of an inch (25 μm). The valve block will be mounted to the machine's frame with a *three point* pattern to avoid distorting the valve block and jamming the valve's sensitive components.

The spool position may be actuated by mechanical levers, hydraulic *pilot* pressure, or solenoids which push the spool left or right. A seal allows part of the spool to protrude outside the housing, where it is accessible to the actuator.

The main valve block is usually a stack of *off the shelf* directional control valves chosen by flow capacity and performance. Some valves are designed to be proportional (flow rate proportional to valve position), while others may be simply on-off.

The control valve is one of the most expensive and sensitive parts of a hydraulic circuit.

- Pressure relief valves are used in several places in hydraulic machinery; on the return circuit to maintain a small amount of pressure for brakes, pilot lines, etc... On hydraulic cylinders, to prevent overloading and hydraulic line/seal rupture. On the hydraulic reservoir, to maintain a small positive pressure which excludes moisture and contamination.
- Pressure regulators reduce the supply pressure of hydraulic fluids as needed for various circuits.
- Sequence valves control the sequence of hydraulic circuits; to ensure that one hydraulic cylinder is fully extended before another starts its stroke, for example.
- Shuttle valves provide a logical or function.
- Check valves are one-way valves, allowing an accumulator to charge and maintain its pressure after the machine is turned off, for example.
- Pilot controlled Check valves are one-way valve that can be opened (for both directions) by a foreign pressure signal. For instance if the load should not be held by the check valve anymore. Often the foreign pressure comes from the other pipe that is connected to the motor or cylinder.
- Counterbalance valves are in fact a special type of pilot controlled check valve. Whereas the check valve is open or closed, the counterbalance valve acts a bit like a pilot controlled flow control.
- Cartridge valves are in fact the inner part of a check valve; they are *off the shelf* components with a standardized envelope, making them easy to populate a proprietary valve block. They are available in many configurations; on/off, proportional, pressure relief, etc. They generally screw into a valve block

and are electrically controlled to provide logic and automated functions.

- Hydraulic fuses are in-line safety devices designed to automatically seal off a hydraulic line if pressure becomes too low, or safely vent fluid if pressure becomes too high.
- Auxiliary valves in complex hydraulic systems may have auxiliary valve blocks to handle various duties unseen to the operator, such as accumulator charging, cooling fan operation, air conditioning power, etc. They are usually custom valves designed for the particular machine, and may consist of a metal block with ports and channels drilled. Cartridge valves are threaded into the ports and may be electrically controlled by switches or a microprocessor to route fluid power as needed.

Actuators

- Hydraulic cylinder
- Swashplates are used in 'hydraulic motors' requiring highly accurate control and also in 'no stop' continuous (360°) precision positioning mechanisms. These are frequently driven by several hydraulic pistons acting in sequence.
- Hydraulic motor (a pump plumbed in reverse)
- Hydrostatic transmission
- Brakes

Reservoir

The hydraulic fluid reservoir holds excess hydraulic fluid to accommodate volume changes from: cylinder extension and contraction, temperature driven expansion and contraction, and leaks. The reservoir is also designed to aid in separation of air from the fluid and also work as a heat accumulator to cover losses in the system when peak power is used. Design engineers are always pressured to reduce the size of hydraulic reservoirs, while equipment operators always appreciate larger reservoirs. Reservoirs can also help separate dirt and other particulate from the oil, as the particulate will generally settle to the bottom of the tank. Some designs include dynamic flow channels on the fluid's return path that allow for a smaller reservoir.

Accumulators

Accumulators are a common part of hydraulic machinery. Their function is to store energy by using pressurized gas. One type is a tube with a floating piston. On one side of the piston is a charge of pressurized gas, and on the other side is the fluid. Bladders are used in other designs. Reservoirs store a system's fluid.

Examples of accumulator uses are backup power for steering or brakes, or to act as a shock absorber for the hydraulic circuit.

Hydraulic Fluid

Also known as *tractor fluid,* hydraulic fluid is the life of the hydraulic circuit. It is usually petroleum oil with various additives. Some hydraulic machines require fire resistant fluids, depending on their applications. In some factories where food is prepared, either an edible oil or water is used as a working fluid for health and safety reasons. In addition to transferring energy, hydraulic fluid needs to lubricate components, suspend contaminants and metal filings for transport to the filter, and to function well to several hundred degrees Fahrenheit or Celsius.

Filters

Filters are an important part of hydraulic systems. Metal particles are continually produced by mechanical components and need to be removed along with other contaminants.

Filters may be positioned in many locations. The filter may be located between the reservoir and the pump intake. Blockage of the filter will cause cavitation and possibly failure of the pump. Sometimes the filter is located between the pump and the control valves. This arrangement is more expensive, since the filter housing is pressurized, but eliminates cavitation problems and protects the control valve from pump failures. The third common filter location is just before the return line enters the reservoir. This location is relatively insensitive to blockage and does not require a pressurized housing, but contaminants that enter the reservoir from external sources are not filtered until passing through the system at least once.

Tubes, Pipes and Hoses

Hydraulic tubes are seamless steel precision pipes, specially manufactured for hydraulics. The tubes have standard sizes for different pressure ranges, with standard diameters up to 100 mm. The tubes are supplied by manufacturers in lengths of 6 m, cleaned, oiled and plugged. The tubes are interconnected by different types of flanges (especially for the larger sizes and pressures), welding cones/ nipples (with o-ring seal), several types of flare connection and by cut-rings. In larger sizes, hydraulic pipes are used. Direct joining of tubes by welding is not acceptable since the interior cannot be inspected.

Hydraulic pipe is used in case standard hydraulic tubes are not available. Generally these are used for low pressure. They can be connected by threaded connections, but usually by welds. Because of the larger diameters the pipe can usually be inspected internally after welding. Black pipe is non-galvanized and suitable for welding.

Hydraulic hose is graded by pressure, temperature, and fluid compatibility. Hoses are used when pipes or tubes can not be used, usually to provide flexibility for machine operation or maintenance. The hose is built up with rubber and steel layers. A rubber interior is surrounded by multiple layers of woven wire and rubber. The exterior is designed for abrasion resistance. The bend radius of hydraulic hose is carefully designed into the machine, since hose failures can be deadly, and violating the hose's minimum bend radius will cause failure. Hydraulic hoses generally have steel fittings swaged on the ends. The weakest part of the high pressure hose is the connection of the hose to the fitting. Another disadvantage of hoses is the shorter life of rubber which requires periodic replacement, usually at five to seven year intervals. Tubes and pipes for hydraulic applications are internally oiled before the system is commissioned. Usually steel piping is painted outside. Where flare and other couplings are used, the paint is removed under the nut, and is a location where corrosion can begin. For this reason, in marine applications most piping is stainless steel.

Seals, Fittings and Connections

In general, valves, cylinders and pumps have female threaded bosses for the fluid connection, and hoses have female ends with captive nuts. A male-male fitting is chosen to connect the two. Many standardized systems are in use.

Fittings serve several purposes;

1. To bridge different standards; O-ring boss to JIC, or pipe threads to face seal, for example.
2. To allow proper orientation of components, a 90°, 45°, straight, or swivel fitting is chosen as needed. They are designed to be positioned in the correct orientation and then tightened.
3. To incorporate bulkhead hardware.
4. A *quick disconnect* fitting may be added to a machine without modification of hoses or valves

A typical piece of heavy equipment may have thousands of sealed connection points and several different types:

- Pipe fittings, the fitting is screwed in until tight, difficult to orient an angled fitting correctly without over or under tightening.
- O-ring boss, the fitting is screwed into a boss and orientated as needed, an additional nut tightens the fitting, washer and o-ring in place.
- Flare fittings, are metal to metal compression seals deformed with a cone nut and pressed into a flare mating.
- Face seal, metal flanges with a groove and o-ring are fastened together.
- Beam seals are costly metal to metal seals used primarily in aircraft.
- Swaged seals, tubes are connected with fittings that are swaged permanently in place. Primarily used in aircraft.

Elastomeric seals (O-ring boss and face seal) are the most common types of seals in heavy equipment and are capable of reliably sealing 6000+ psi (40+ MPa) of fluid pressure.

Basic Calculations

Hydraulic power is defined as flow times pressure. The hydraulic power supplied by a pump:

Power = (P x Q) ÷ 600

where power is in kilowatts [kW], P pressure in bars, and Q is the flow in liters per minute. For example, a pump delivers 180 lit/min and the pressure equals 250 bar, therefore the power of the pump is 75 kW.

When calculating the power input to the pump, the total pump efficiency η_{total} must be included. This efficiency is the product of volumetric efficiency, η_{vol} and the hydromechanical efficiency, η_{hm}. Power input = Power output ÷ η_{total}. The average for axial piston pumps, $\eta_{total} = 0.87$. In the example the power source, for example a diesel engine or an electric motor, must be capable of delivering at least 75 ÷ 0.87 = 86 [kW]. The hydraulic motors and cylinders that the pump supplies with hydraulic power also have efficiencies and the total system efficiency (without including the pressure drop in the hydraulic pipes and valves) will end up at approx. 0.75. Cylinders normally have a total efficiency around 0.95 while hydraulic axial piston motors 0.87, the same as the pump. In general the power loss in a hydraulic energy transmission is thus around 25% or more at ideal viscosity range 25-35 [St]. Calculation of the required max. power output for the diesel engine, rough estimation:

(1) Check the max. powerpoint, i.e. the point where pressure times flow reach the max. value.

(2) $E_{diesel} = (P_{max} \cdot Q_{tot}) \div \eta$.

Q_{tot} = calculate with the theoretical pump flow for the consumers not including leakages at max. power point.

P_{max} = actual pump pressure at max. power point.

Note: η is the total efficiency = (output mechanical power ÷ input mechanical power). For rough estimations, $\eta = 0.75$. Add 10-20% (depends on the application) to this power value.

(3) Calculate the required pumpdisplacement from required max. sum of flow for the consumers in worst case and the diesel engine rpm in this point. The max. flow can differ from the flow used for calculation of the diesel engine power. Pump volumetric efficiency average, piston pumps: $\eta_{vol} = 0.93$.

Pumpdisplacement $V_{pump} = Q_{tot} \div n_{diesel} \chi 0.93$.

(4) Calculation of prel. cooler capacity: Heat dissipation from hydraulic oil tanks, valves, pipes and hydraulic components is less than a few percent in standard mobile equipment and the cooler capacity must include some margins. Minimum cooler capacity, $E_{cooler} = 0.25E_{diesel}$

At least 25% of the input power must be dissipated by the cooler when peak power is utilised for long periods. In normal case however, the peak power is used for only short periods, thus the actual cooler capacity required might be considerably less. The oil volume in the hydraulic tank is also acting as a heat accumulator when peak power is used. The system efficiency is very much dependent on the type of hydraulic work tool equipment, the hydraulic pumps and motors used and power input to the hydraulics may vary a lot.

Each circuit must be evaluated and the load cycle estimated. New or modified systems must always be tested in practical work, covering all possible load cycles. An easy way of measuring the actual average power loss in the system is to equip the machine with a test cooler and measure the oil temperature at cooler inlet, oil temperature at cooler outlet and the oil flow through the cooler, when the machine is in normal operating mode.

From these figures the test cooler power dissipation can be calculated and this is equal to the power loss when temperatures are stabilized. From this test the actual required cooler can be calculated to reach specified oil temperature in the oil tank. One problem can be to assemble the measuring equipment inline, especially the oil flow metre.

History

Joseph Bramah patented the hydraulic press in 1795. While working at Bramah's shop, Henry Maudslay suggested a cup leather packing. Because it produced superior results, the hydraulic press eventually displaced the steam hammer from metal forging.

To supply small scale power that was impractical for individual steam engines, central station hydraulic systems were developed. Hydraulic power was used to operate cranes and other machinery in British ports and elsewhere in Europe. The largest hydraulic system was in London. Hydraulic power was used extensively in Bessemer steel production. Hydraulic power was also used for elevators, to operate canal locks and rotating sections of bridges. Some of these systems remained in use well into the twentieth century.

Harry Franklin Vickers was called the "Father of Industrial Hydraulics" by ASME.

Automatic Transmission

An automatic transmission (also called automatic gearbox) is a type of motor vehcle transmission that can automatically change gear ratios as the vehicle moves, freeing the driver from having to shift gears manually. Most automatic transmissions have a defined set of gear ranges, often with a parking pawl feature that locks the output shaft of the transmission stroke face.

Similar but larger devices are also used for heavy-duty commercial and industrial vehicles and equipment. Some machines with limited speed ranges or fixed engine speeds, such as some forklifts and lawn mowers, only use a torque converter to provide a variable gearing of the engine to the wheels.

Besides automatics, there are also other types of automated transmissions such as a continuously variable transmission (CVT) and semi-automatic transmissions, that free the driver from having to shift gears manually, by using the transmission's computer to change gear, if for example the driver were redlining the engine. Despite superficial similarity to other transmissions, automatic transmissions differ significantly in internal operation and driver's feel from semi-automatics and CVTs. An automatic uses a torque converter instead of a clutch to manage the connection between the transmission gearing and the engine. In contrast, a CVT uses a belt or other torque transmission scheme to allow an "infinite" number of gear ratios instead of a fixed number of gear ratios. A semi-automatic retains a clutch like a manual transmission, but controls the clutch through electrohydraulic means.

A conventional manual transmission is frequently the base equipment in a car, with the option being an automated transmission such as a conventional automatic, semi-automatic, or CVT. The ability to shift gears manually, often via paddle shifters, can also be found on certain automated transmissions (manumatics such as Tiptronic), semi-automatics (BMW SMG), and CVTs (such as Lineartronic).

The first automatic transmission was invented in 1921 by Alfred Horner Munro of Regina, Saskatchewan, Canada, and patented under Canadian patent CA 235757 in 1923. (Munro obtained UK patent GB215669 215,669 for his invention in 1924 and US patent 1,613,525 on 4 January 1927). Being a steam engineer, Munro designed his device to use compressed air rather than hydraulic fluid, and so it lacked power and never found commercial application. The first automatic transmissions using hydraulic fluid were developed by General Motors during the 1930s and introduced in the 1940 Oldsmobile as the "Hydra-Matic" transmission. They were incorporated into GM-built tanks during World War II and, after the war, GM marketed them as being "battle-tested".

Comparison with Manual Transmission

Most cars sold in North America since the 1950s have been available with an automatic transmission. Conversely, in Europe a manual gearbox is standard, with 20% of drivers opting for an automatic transmission. In some Asian markets and in Australia, automatic transmissions have become very popular since the 1990s.

Vehicles equipped with automatic transmissions are not as complex to drive. Consequently, in some jurisdictions, drivers who have passed their driving test in a vehicle with an automatic transmission will not be licensed to drive a manual transmission vehicle. Conversely, a manual license will allow the driver to drive both manual and automatic vehicles. Examples of driving license restrictions are Croatia, Dominican Republic, Israel, United Kingdom, Brazil, some states in Australia, France, Portugal, Latvia, Lebanon, Lithuania, Ireland, Belgium, Germany, the Netherlands, Sweden, Spain, Austria, Norway, Hungary, South Africa, Trinidad and Tobago, Belize, Japan, China,

Hong Kong, Macau, Mauritius, South Korea, Romania, Singapore, Philippines, United Arab Emirates, India, Estonia, Finland, Saudi Arabia (in March 2012), Switzerland, Slovenia, Republic of Ireland and New Zealand (restricted licence only).

Automatic Transmission Modes

Conventionally, in order to select the transmission operating mode, the driver moves a selection lever located either on the steering column or on the floor (as with a manual on the floor, except that most automatic selectors on the floor do not move in the same type of pattern as a manual lever; most automatic levers only move vertically). In order to select modes, or to manually select specific gear ratios, the driver must push a button in (called the shift lock button) or pull the handle (only on column mounted shifters) out.

Some vehicles position selector buttons for each mode on the cockpit instead, freeing up space on the central console. Vehicles conforming to US Government standards must have the modes ordered P-R-N-D-L (left to right, top to bottom, or clockwise). Prior to this, quadrant-selected automatic transmissions often used a P-N-D-L-R layout, or similar. Such a pattern led to a number of deaths and injuries owing to driver error causing unintentional gear selection, as well as the danger of having a selector (when worn) jump into Reverse from Low gear during engine braking maneuvers.

Automatic transmissions have various modes depending on the model and make of the transmission. Some of the common modes include:

Park (P)

This selection mechanically locks the output shaft of transmission, restricting the vehicle from moving in any direction. A parking pawl prevents the transmission from rotating, and therefore the vehicle from moving, but the vehicle's driven wheels may still rotate individually (because of the differential), as well as the non-driven roadwheels may still rotate freely. For this reason, it is recommended to use the hand brake (*parking brake*) because this actually locks (in most cases) the wheels and prevents them from moving. This also increases the life of the transmission and the park pin mechanism, because parking on an incline with the transmission in park without the parking brake engaged will cause undue stress on the parking pin. A hand brake should also prevent the car from moving if a worn selector accidentally drops into reverse gear while idleing.

It should be noted that locking the transmission output shaft with park does not definitivly lock the driving wheels. If one driving wheel has little vertical load it will tend to slip, and will rotate in the opposite direction to the more heavily loaded non-slipping wheel. Only a parking brake can be relied upon to positively lock both of the parking-braked wheels. (This is not the case with certain 1950's Chrysler products that carried their parking brake on the transmission tailshaft, a defect compounded by the provision of a bumper jack.) It is typical of front-wheel-drive vehicles for the parking brake to lock the rear (non-driving) wheels, so use of both the parking brake and the transmission park lock provides the greatest security against unintended movement on slopes. Buses in Hong Kong *(Kowloon Motor Bus)* do not have Parking mode (P), instead, they are using a single lever to prevent the whole bus from moving.

A car should be allowed to come to a complete stop before setting the transmission into park to prevent damage. Usually, Park (P) is one of only two selections in which the car's engine can be started, the other being Neutral (N) (Buses in Hong Kong mentioned above has to be in Neutral mode (N) ONLY before the engine can be started). In many modern cars and trucks, the driver must have the foot brake applied before the transmission can be taken out of park. The Park position is omitted on buses/coaches (and some road tractors) with automatic transmission (on which a parking pawl is not practical), which must be placed in neutral with the air-operated parking brakes set. Advice is given in some owner's manuals that if the vehicle is parked on a steep slope using the park lock only, it may not be possible to release the park lock (move the selector lever out of P). Another vehicle may be required to push the stuck vehicle uphill slightly to remove the loading on the park lock pawl.

Most automobiles require P or N to be set on the selector lever before the engine can be started. This is typically achieved via a normally-open inhibitor switch (sometimes called a "neutral safety switch") wired in series with the starter motor engagement circuit, which is closed when P or N is selected, completing the circuit (when the key is turned to the start position), along with any other safety devices which may be present on newer cars (such as a foot-brake application).

Reverse (R)

This engages reverse gear within the transmission, permitting the vehicle to be driven backward, and operates a switch to turn on the backup lights for improved visibility (the switch may also activate a beeper on delivery trucks or other large vehicles to warn other drivers and nearby pedestrians of the driver's reverse movement). To select reverse in most transmissions, the driver must come to a complete stop, depress the shift lock button (or move the shift lever toward the driver in a column shifter, or move the shifter sideways along a notched channel in a console shifter) and select reverse. Not coming to a complete stop may cause severe damage to the transmission.

Some modern automatic transmissions have a safety mechanism in place, which does, to some extent, prevent (but not completely avoid) inadvertently putting the car in reverse when the vehicle is moving forward; such a mechanism may consist of a solenoid-controlled physical barrier on either side of the Reverse position, electronically engaged by a switch on the brake pedal. Therefore, the brake pedal needs to be depressed in order to allow the selection of reverse. Some electronic transmissions prevent or delay engagement of reverse gear altogether while the car is moving.

Some shifters with a shift button allow the driver to freely move the shifter from R to N or D, or simply moving the shifter to N or D without actually depressing the button. However, the driver cannot shift back to R without depressing the shift button, to prevent accidental shifting, especially at high speeds, which could damage the transmission.

Neutral / No Gear (N)

This disengages all gear trains within the transmission, effectively disconnecting the transmission from the driven wheels, allowing the vehicle to coast freely under its own weight and gain momentum without the motive force from the engine. Coasting in idle down long grades (where law permits) should be avoided, though, as the transmission's lubrication pump is driven by non-idle engine RPMs. Similarly, emergency towing with an automatic transmission in neutral should be a last resort. Manufacturers understand emergency situations and list limitations of towing a vehicle in neutral (usually not to exceed 55 mph and 50 miles). This is the only other selection in which the vehicle's engine may be started.

Drive (D)

This position allows the transmission to engage the full range of available forward gear ratios, allowing the vehicle to move forward and accelerate through its range of gears. The number of gear ratios within the transmission depends on the model, but they initially ranged from three (predominant before the 1990s), to four and five speeds (losing popularity to six-speed autos, though still favoured by Chrysler and Honda/Acura). Six-speed automatic transmissions are probably the most common offering in cars and trucks from 2010 in carmakers as Toyota, GM and Ford. However, seven-speed automatics are becoming available in some high-performance production luxury cars (found in Mercedes 7G gearbox, Infiniti), as are eight-speed autos in models from 2006 introduced by Aisin Seiki Co. in Lexus, ZF and Hyundai Motor Company. From 2013 are available nine speeds transmissions produced by ZF and Mercedes 9G.

Overdrive ('D', 'OD', or a boxed [D] or the absence of an illuminated 'O/D OFF')

This mode is used in some transmissions to allow early computer-controlled transmissions to engage the automatic overdrive. In these transmissions, Drive (D) locks the automatic overdrive off, but is identical otherwise. OD (Overdrive) in these cars is engaged under steady speeds or low acceleration at approximately 35–45 mph (56–72 km/h). Under hard acceleration or below 35–45 mph (56–72 km/h), the transmission will automatically downshift. Other vehicles with this selector (example light trucks) will not only disable up-shift to the overdrive gear, but keep the remaining available gears continuously engaged to the engine for use of compression braking. Verify the behaviour of this switch and consider the benefits of reduced friction brake use when city driving where speeds typically do not necessitate the overdrive gear.

Third (3)

This mode limits the transmission to the first three gear ratios, or sometimes locks the transmission in third gear. This can be used to climb or going down hill. Some vehicles will automatically shift up out of third gear in this mode if a certain revolutions per minute (RPM) range is reached in order to prevent engine damage. This gear is also recommended while towing a trailer.

Second (2 or S)

This mode limits the transmission to the first two gear ratios, or locks the transmission in second gear on Ford, Kia, and Honda models. This can be used to drive in adverse conditions such as snow and ice, as well as climbing or going down hills in winter. It is usually recommended to use second gear for starting on snow and ice, and use of this position enables this with an automatic transmission. Some vehicles will automatically shift up out of second gear in this mode if a certain RPM range is reached in order to prevent engine damage.

Although traditionally considered second gear, there are other names used. Chrysler models with a three-speed automatic since the late 1980s have called this gear 3 while using the traditional names for *Drive* and *Low*. Oldsmobile has called second gear as the 'Super' range — which was first used on their 4-speed Hydramatic transmissions, although the use of this term continued until the early 1980s when GM's Turbo Hydramatic automatic transmissions were standardized by all of their divisions years after the 4-speed Hydramatic was discontinued.

First (1 or L [Low])

This mode locks the transmission in first gear only. In older vehicles, it will not change to any other gear range. Some vehicles will automatically shift up out of first gear in this mode if a certain RPM range is reached in order to prevent engine damage. This, like second, can be used during the winter season, for towing, or for downhill driving to increase the engine braking effect. As well as the above modes there are also other modes, dependent on the manufacturer and model. Some examples include:

D5

In Hondas and Acuras equipped with five-speed automatic transmissions, this mode is used commonly for highway use (as stated in the manual), and uses all five forward gears.

D4

This mode is also found in Honda and Acura four or five-speed automatics, and only uses the first four gear ratios. According to the manual, it is used for stop-and-go traffic, such as city driving.

D3 or 3

This mode is found in Honda, Acura, Volkswagen and Pontiac four-speed automatics and only uses the first three gear ratios. According to the manual, it is used for stop-and-go traffic, such as city driving.

D2 and D1

These modes are found on older Ford transmissions (C6, etc.). In D1, all three gears are used, whereas in D2 the car starts in second gear and upshifts to third.

S or Sport

This is commonly described as Sport mode. It operates in an identical manner as "D" mode, except that the upshifts change much higher up the engine's rev range. This has the effect on maximising all the available engine output, and therefore enhances the performance of the vehicle, particularly during acceleration. This mode will also downchange much higher up the rev range compared to "D" mode, maximising the effects of engine braking. This mode will have a detrimental effect on fuel economy. Hyundai has a Norm/Power switch next to the gearshift for this purpose on the *Tiburon*.

Some early GMs equipped with HYDRA-MATIC transmissions used (S) to indicate Second gear, being the same as the 2 position on a Chrysler, shifting between only first and second gears. This would have been recommended for use on steep grades, or slippery roads like dirt, or ice, and limited to speeds under 40 mph. (L) was used in some early GMs to indicate (L)ow gear, being the same as the 2 position on a Chrysler, locking the transmission into first gear. This would have been recommended for use on steep grades, or slippery roads like dirt, or ice, and limited to speeds under 15 mph.

+ –, and M

This is for the Manual mode selection of gears in certain automatics, such as Porsche's Tiptronic and Honda's StepTronic. The M feature can also be found in Chrysler and General Motors products such as the Dodge Magnum, Journey, and Pontiac G6, Mazda products such as the Mazda 3, Mazda6, and the CX-7, as well as Toyota's Camry, Corolla, Fortuner, Previa and Innova. Mitsubishi and some Audi models (Audi TT), meanwhile do not have the M, and instead have the + and -, which is separated from the rest of the shift modes; the same is true for some Peugeot products like Peugeot 206. Meanwhile, the driver can shift up and down at will by toggling the (console mounted) shift lever similar to a semi-automatic transmission. This mode may be engaged either through a selector/position or by actually changing the gears (e.g., tipping the gear-down paddles mounted near the driver's fingers on the steering wheel).

Winter (W)

In some Volvo, Mercedes-Benz, BMW and General Motors Europe models, a winter mode can be engaged so that second gear is selected instead of first when pulling away from stationary, to reduce the likelihood of loss of traction due to wheel spin on snow or ice. On GM cars, this was D2 in the 1950s, and is Second Gear Start after 1990. On Ford, Kia, and Honda automatics, this feature can be accessed by moving the gear selector to 2 to start, then taking your foot off the accelerator while selecting D once the car is moving.

Brake (B)

A mode selectable on some Toyota models. In non-hybrid cars, this mode lets the engine do compression braking, also known as engine braking, typically when encountering a steep downhill. Instead of engaging the brakes, the engine in a non-hybrid car switches to a lower gear and slows down the spinning tires. The engine holds the car back, instead of the brakes slowing it down. GM called this "HR" ("hill retarder") and "GR" ("grade retarder") in the 1950s. For hybrid cars, this mode converts the electric motor into a generator for the battery (Regenerative Braking). It is not the same as downshifting in a non-hybrid car, but it has the same effect in slowing the car without using the brakes.

Hydraulic Automatic Transmissions

The predominant form of automatic transmission is hydraulically operated; using a fluid coupling or torque converter, and a set of planetary gearsets to provide a range of gear ratios.

Parts and Operation

A hydraulic automatic transmission consists of the following parts:

- *Torque converter:* A type of fluid coupling, hydraulically connecting the engine to the transmission. It takes the place of a mechanical clutch, allowing the transmission to stay in gear and the engine to remain running while the vehicle is stationary, without stalling. A torque converter differs from a fluid coupling, in that it provides a variable amount of torque multiplication at low engine speeds, increasing breakaway acceleration. This is accomplished with a third member in the coupling assembly known as the stator, and by altering the shapes of the vanes inside the coupling in such a way as to curve the fluid's path into the stator. The stator captures the kinetic energy of the transmission fluid, in effect using the leftover force of it to enhance torque multiplication.
- *Pump*: Not to be confused with the impeller inside the torque converter, the pump is typically a gear pump mounted between the torque converter and the planetary gearset. It draws transmission fluid from a sump and pressurizes it, which is

needed for transmission components to operate. The input for the pump is connected to the torque converter housing, which in turn is bolted to the engine's flywheel, so the pump provides pressure whenever the engine is running and there is enough transmission fluid. Early automatic transmissions also had a rear pump, allowing push-starting.

- *Planetary gearset*: A compound epicyclic planetary gearset, whose bands and clutches are actuated by hydraulic servos controlled by the valve body, providing two or more gear ratios. (Not part of some manufacturers transmissions during some eras, Honda being but one).
- *Clutches and bands:* to effect gear changes, one of two types of clutches or bands are used to hold a particular member of the planetary gearset motionless, while allowing another member to rotate, thereby transmitting torque and producing gear reductions or overdrive ratios. These clutches are actuated by the valve body, their sequence controlled by the transmission's internal programming. Principally, a type of device known as a sprag or roller clutch is used for routine upshifts/downshifts. Operating much as a ratchet, it transmits torque only in one direction, free-wheeling or "overrunning" in the other. The advantage of this type of clutch is that it eliminates the sensitivity of timing a simultaneous clutch release/apply on two planetaries, simply "taking up" the drivetrain load when actuated, and releasing automatically when the next gear's sprag clutch assumes the torque transfer. The bands come into play for manually selected gears, such as low range or reverse, and operate on the planetary drum's circumference. Bands are not applied when drive/overdrive range is selected, the torque being transmitted by the sprag clutches instead. Bands are used for braking; the GM Turbo-Hydramatics incorporated this..
- *Valve body*: hydraulic control centre that receives pressurized fluid from the *main pump* operated by the fluid coupling/ torque converter. The pressure coming from this pump is regulated and used to run a network of spring-loaded valves, check balls and servo pistons. The valves use the pump pressure and the pressure from a centrifugal governor on the output side (as well as hydraulic signals from the range selector valves and the *throttle valve* or *modulator*) to control which ratio is selected on the gearset; as the vehicle and engine

change speed, the difference between the pressures changes, causing different sets of valves to open and close. The hydraulic pressure controlled by these valves drives the various clutch and brake band actuators, thereby controlling the operation of the planetary gearset to select the optimum gear ratio for the current operating conditions. However, in many modern automatic transmissions, the valves are controlled by electro-mechanical servos which are controlled by the electronic engine control unit (ECU) or a separate transmission control unit (TCU, also known as transmission control module (TCM).

- *Hydraulic & lubricating oil*: called automatic transmission fluid (ATF), this component of the transmission provides lubrication, corrosion prevention, and a hydraulic medium to convey mechanical power (for the operation of the transmission). Primarily made from refined petroleum, and processed to provide properties that promote smooth power transmission and increase service life, the ATF is one of the few parts of the automatic transmission that needs routine service as the vehicle ages.

The multitude of parts, along with the complex design of the valve body, originally made hydraulic automatic transmissions much more complicated (and expensive) to build and repair than manual transmissions. In most cars (except US family, luxury, sport-utility vehicle, and minivan models) they have usually been extra-cost options for this reason. Mass manufacturing and decades of improvement have reduced this cost gap.

Energy Efficiency

Hydraulic automatic transmissions are almost always less energy efficient than manual transmissions due mainly to viscous and pumping losses, both in the torque converter and the hydraulic actuators. A relatively small amount of energy is required to pressurize the hydraulic control system, which uses fluid pressure to determine the correct shifting patterns and operate the various automatic clutch mechanisms.

Manual transmissions use a mechanical clutch to transmit torque, rather than a torque converter, thus avoiding the primary source of loss in an automatic transmission. Manual transmissions also avoid the power requirement of the hydraulic control system, by relying on the human muscle power of the vehicle operator to disengage the clutch and actuate the gear levers, and the mental power of the operator to make appropriate gear ratio selections. Thus the manual transmission requires very little engine power to function, with the main power consumption due to drag from the gear train being immersed in the lubricating oil of the gearbox.

The on-road acceleration of an automatic transmission can occasionally exceed that of an otherwise identical vehicle equipped with a manual transmission in turbocharged diesel applications. Turbo-boost is normally lost between gear changes in a manual whereas in an automatic the accelerator pedal can remain fully depressed. This however is still largely dependent upon the number and optimal spacing of gear ratios for each unit, and whether or not the elimination of spooldown/accelerator lift off represent a significant enough gain to counter the slightly higher power consumption of the automatic transmission itself.

History and Improvements

Modern automatic transmissions can trace their origins to an early "horseless carriage" gearbox that was developed in 1904 by the Sturtevant brothers of Boston, Massachusetts. This unit had two forward speeds, the ratio change being brought about by flyweights that were driven by the engine. At higher engine speeds, high gear was engaged. As the vehicle slowed down and engine RPM decreased, the gearbox would shift back to low. Unfortunately, the metallurgy of the time wasn't up to the task, and owing to the abruptness of the gear change, the transmission would often fail without warning.

The next significant phase in the automatic transmission's development occurred in 1908 with the introduction of Henry Ford's remarkable Model T. The Model T, in addition to being cheap and reliable by the standards of the day, featured a simple, two speed plus reverse planetary transmission whose operation was manually controlled by the driver using pedals. The pedals actuated the transmission's friction elements (bands and clutches) to select the desired gear. In some respects, this type of transmission was less demanding of the driver's skills than the contemporary, unsynchronized manual transmission, but still required that the driver know when to make a shift, as well as how to get the car off to a smooth start.

In 1934, both REO and General Motors developed semi-automatic transmissions that were less difficult to operate than a fully manual unit. These designs, however, continued to use a clutch to engage the engine with the transmission. The General Motors unit, dubbed the "Automatic Safety Transmission," was notable in that it employed a power-shifting planetary gearbox that was hydraulically controlled and was sensitive to road speed, anticipating future development. Parallel to the development in the 1930s of an automatically shifting gearbox was Chrysler's work on adapting the fluid coupling to automotive use. Invented early in the 20th century, the fluid coupling was the answer to the question of how to avoid stalling the engine when the vehicle was stopped with the transmission in gear. Chrysler itself never used the fluid coupling with any of its automatic transmissions, but did use it in conjunction with a hybrid manual transmission called "Fluid Drive" (the similar Hy-Drive used a torque converter). These developments in automatic gearbox and fluid coupling technology eventually culminated in the introduction in 1939 of the General Motors Hydra-Matic, the world's first mass-produced automatic transmission.

Available as an option on 1940 Oldsmobiles and later Cadillacs, the Hydra-Matic combined a fluid coupling with three hydraulically-controlled planetary gearsets to produce four forward speeds plus reverse. The transmission was sensitive to engine throttle position and road speed, producing fully automatic up- and down-shifting that varied according to operating conditions.

The Hydra-Matic was subsequently adopted by Cadillac and Pontiac, and was sold to various other automakers, including Bentley, Hudson, Kaiser, Nash, and Rolls-Royce. It also found use during World War II in some military vehicles. From 1950-1954, Lincoln cars were also available with the Hydra-Matic. Mercedes-Benz subsequently devised a four-speed fluid coupling transmission that was similar in principle to the Hydra-Matic, but of a different design.

Interestingly, the original Hydra-Matic incorporated two features which are widely emulated in today's transmissions. The Hydra-Matic's ratio spread through the four gears produced excellent "step-off" and acceleration in first, good spacing of intermediate gears, and the effect of an overdrive in fourth, by virtue of the low numerical rear axle ratio used in the vehicles of the time. In addition, in third and fourth gear, the fluid coupling only handled a portion of the engine's torque, resulting in a high degree of efficiency. In this respect, the transmission's behaviour was similar to modern units incorporating a lock-up torque converter.

In 1956, GM introduced the "Jetaway" Hydra-Matic, which was different in design than the older model. Addressing the issue of shift quality, which was an ongoing problem with the original Hydra-Matic, the new transmission utilised two fluid couplings, the primary one that linked the transmission to the engine, and a secondary one that replaced the clutch assembly that controlled the forward gearset in the original. The result was much smoother shifting, especially from first to second gear, but with a loss in efficiency and an increase in complexity. Another innovation for this new style Hydra-Matic was the appearance of a Park position on the selector. The original Hydra-Matic, which continued in production until the mid-1960s, still used the Reverse position for parking pawl engagement.

The first torque converter automatic, Buick's Dynaflow, was introduced for the 1948 model year. It was followed by Packard's Ultramatic in mid-1949 and Chevrolet's Powerglide for the 1950 model year. Each of these transmissions had only two forward speeds, relying on the converter for additional torque multiplication. In the early 1950s, BorgWarner developed a series of three-speed torque converter automatics for American Motors, Ford Motor Company, Studebaker, and several other manufacturers in the US and other countries. Chrysler was late in developing its own true automatic, introducing the two-speed torque converter PowerFlite in 1953, and the three-speed TorqueFlite in 1956. The latter was the first to utilise the Simpson compound planetary gearset.

General Motors produced multiple-turbine torque converters from 1954 to 1961. These included the Twin-Turbine Dynaflow and the triple-turbine Turboglide transmissions. The shifting took place in the torque converter, rather than through pressure valves and changes in planetary gear connections. Each turbine was connected to the drive shaft through a different gear train. These phased from one ratio to another according to demand, rather than shifting. The Turboglide actually had two speed ratios in reverse, with one of the turbines rotating backwards.

By the late 1960s, most of the fluid-coupling four-speed and two-speed transmissions had disappeared in favour of three-speed units with torque converters. Also around this time, whale oil was removed from automatic transmission fluid. By the early 1980s, these were being supplemented and eventually replaced by overdrive-equipped transmissions providing four or more forward speeds. Many transmissions also adopted the lock-up torque converter (a mechanical clutch locking the torque converter pump and turbine together to eliminate slip at cruising speed) to improve fuel economy.

As computerised engine control units (ECUs) became more capable, much of the logic built into the transmission's valve body was offloaded to the ECU. (Some manufacturers use a separate computer dedicated to the transmission, but sharing information with the engine management computer.) In this case, solenoids turned on and off by the computer control shift patterns and gear ratios, rather than the spring-loaded valves in the valve body. This allows for more precise control of shift points, shift quality, lower shift times, and (on some newer cars) semi-automatic control, where the driver tells the computer when to shift. The result is an impressive combination of efficiency and smoothness. Some computers even identify the driver's style and adapt to best suit it.

ZF Friedrichshafen and BMW were responsible for introducing the first six-speed (the ZF 6HP26 in the 2002 BMW E65 7-Series). Mercedes-Benz's 7G-Tronic was the first seven-speed in 2003, with Toyota introducing an eight-speed in 2007 on the Lexus LS 460. Derived from the 7G-Tronic, Mercedes-Benz unveiled a semi-automatic transmission with the torque converter replaced with a wet multi clutch called the AMG SPEEDSHIFT MCT.

Automatic Transmission Models

Some of the best known automatic transmission families include:

- General Motors — Dynaflow, Powerglide, Turboglide, "Turbo-Hydramatic" TH350, TH400 and 700R4, 4L60-E, 4L80-E, Holden Trimatic
- Ford: Cruise-O-Matic, C4, CD4E, C6, AOD/AODE, E4OD, ATX, AXOD/AX4S/AX4N
- Cummins 68 RFE (fitted to the Ram diesel segment)
- Chrysler: TorqueFlite 727 and 904, A500, A518, 45RFE, 545RFE
- BorgWarner (later Aisin AW)
- ZF Friedrichshafen automatic transmissions
- Mercedes-Benz transmissions
- Allison Transmission
- Voith Voith Turbo
- Aisin AW; Aisin AW is a Japanese automotive parts supplier, known for its automatic transmissions and navigation systems
- Honda
- Nissan/Jatco
- Volkswagen Group — 01M
- Drivetrain Systems International (DSI) — M93, M97 and M74 4-speeds, M78 and M79 6-speeds

- Hyundai Hyundai Powertech — 4F12, 4F16, 4F23 4-Speeds, 5F25, 5F16, 5F23 5-Speeds, 6F17, 6F26, 6F40 6-Speeds, 8R40, 8R50 8-Speeds, Mini Cooper — Automatic or manual transmission, all models

Automatic transmission families are usually based on Ravigneaux, Lepelletier, or Simpson planetary gearsets. Each uses some arrangement of one or two central sun gears, and a ring gear, with differing arrangements of planet gears that surround the sun and mesh with the ring. An exception to this is the Hondamatic line from Honda, which uses sliding gears on parallel axes like a manual transmission without any planetary gearsets. Although the Honda is quite different from all other automatics, it is also quite different from an automated manual transmission (AMT). Many of the above AMTs exist in modified states, which were created by racing enthusiasts and their mechanics by systematically re-engineering the transmission to achieve higher levels of performance. These are known as "performance transmissions". An example of a manufacturer of high performance transmissions of General Motors and Ford transmissions is Performa Built.

Continuously Variable Transmissions

A fundamentally different type of automatic transmission is the *continuously variable transmission* or *CVT*, which can smoothly and steplessly alter its gear ratio by varying the diameter of a pair of belt or chain-linked pulleys, wheels or cones. Some continuously variable transmissions use a hydrostatic drive — consisting of a variable displacement pump and a hydraulic motor — to transmit power without gears. CVT designs are usually as fuel efficient as manual transmissions in city driving, but early designs lose efficiency as engine speed increases.

A slightly different approach to CVT is the concept of *toroidal CVT* or *infinitely variable transmission* (IVT). These concepts provide zero and reverse gear ratios. Some current hybrid vehicles, notably those of Toyota, Lexus and Ford Motor Company, have an electronically controlled CVT (E-CVT). In this system, the transmission has fixed gears, but the ratio of wheel-speed to engine-speed can be continuously varied by controlling the speed of the third input to a differential using an electric motor-generator.

Manually Controlled Automatic Transmissions

Most automatic transmissions offer the driver a certain amount of manual control over the transmission's shifts (beyond the obvious selection of forward, reverse, or neutral). Those controls take several forms:

Throttle Kickdown: Most automatic transmissions include some means of forcing a downshift into the lowest possible gear ratio if the throttle pedal is fully depressed. In many older designs, kickdown is accomplished by mechanically actuating a valve inside the transmission. Most modern designs use a solenoid-operated valve that is triggered by a switch on the throttle linkage or by the engine control unit (ECU) in response to an abrupt increase in engine power.

Mode Selection: Allows the driver to choose between preset shifting programs. For example, Economy mode saves fuel by upshifting at lower engine speeds, while Sport mode (aka "Power" or "Performance") delays upshifting for maximum acceleration. Some transmission units also have Winter mode, where higher gear ratios are chosen to keep revs as low as possible while on slippery surfaces. The modes also change how the computer responds to throttle input.

Low Gear Ranges: Conventionally, automatic transmissions have selector positions that allow the driver to limit the maximum ratio that the transmission may engage. On older transmissions, this was accomplished by a mechanical lockout in the transmission valve body preventing an upshift until the lockout was disengaged; on computer-controlled transmissions, the same effect is accomplished by firmware. The transmission can still upshift and downshift automatically between the remaining ratios: for example, in the *3* range, a transmission could shift from first to second to third, but not into fourth or higher ratios. Some transmissions will still upshift automatically into the higher ratio if the engine reaches its maximum permissible speed in the selected range.

Manual Controls: Some transmissions have a mode in which the driver has full control of ratio changes (either by moving the selector, or through the use of buttons or paddles), completely overriding the automated function of the hydraulic controller. Such control is particularly useful in cornering, to avoid unwanted upshifts or downshifts that could compromise the vehicle's balance or traction. "Manumatic" shifters, first popularized by Porsche in the 1990s under the trade name Tiptronic, have become a popular option on sports cars and other performance vehicles. With the near-universal prevalence of electronically controlled transmissions, they are comparatively simple and inexpensive, requiring only software changes, and the provision of the actual manual controls for the driver. The amount of true manual control provided is highly variable: some systems will override the driver's selections under certain conditions, generally in the interest of preventing engine damage. Since these gearboxes also have a throttle kickdown switch, it is impossible to fully exploit the engine power at low to medium engine speeds.

Second Gear Takeoff: Some automatics, particularly those fitted to larger capacity or high torque engines, either when "2" is manually selected, or by engaging a winter mode, will start off in second gear instead of first, and then not shift into a higher gear until returned to "D." Also note that as with most American automatic transmissions, selecting "2" using the selection lever will not tell the transmission to be in only 2nd gear; rather, it will simply limit the transmission to 2nd gear after prolonging the duration of 1st gear through higher speeds than normal operation. The 2000–2002 Lincoln LS V8 (the five-speed automatic *without* manumatic capabilities, as opposed to the optional sport package w/ manu-matic 5-speed) started in 2nd gear during most starts both in winter and other seasons by selecting the "D5" transmission selection notch in the shiftgate (for fuel savings), whereas "D4" would always start in 1st gear. This is done to reduce torque multiplication when proceeding forward from a standstill in conditions where traction was limited — on snow- or ice-covered roads, for example.

Some automatic transmissions modified or designed specifically for drag racing may also incorporate a transmission brake, or "trans-brake," as part of a manual valve body. Activated by electrical solenoid control, a trans-brake simultaneously engages the first and reverse gears, locking the transmission and preventing the input shaft from turning. This allows the driver of the car to raise the engine RPM against the resistance of the torque converter, then launch the car by simply releasing the trans-brake switch.

Hydraulic Brake

The hydraulic brake is an arrangement of braking mechanism which uses brake fluid, typically containing ethylene glycol, to transfer pressure from the controlling mechanism to the braking mechanism.

History

Fred Duesenberg originated hydraulic brakes on his 1914 racing cars and Duesenberg was the first automotive marque to use the technology on a passenger car in 1920. This braking system could have earned him a fortune if he had patented it. In 1918 Malcolm Lougheed (who later changed the spelling of his name to Lockheed) developed a hydraulic brake system.

Construction

The most common arrangement of hydraulic brakes for passenger vehicles, motorcycles, scooters, and mopeds, consists of the following:

- Brake pedal or lever
- A pushrod (also called an *actuating rod*)
- A master cylinder assembly containing a piston assembly (made up of either one or two pistons, a return spring, a series of gaskets/ O-rings and a fluid reservoir)
- Reinforced hydraulic lines
- Brake caliper assembly usually consisting of one or two hollow aluminium or chrome-plated steel pistons (called *caliper pistons*), a set of thermally conductive brake pads and a rotor (also called a *brake disc*) or drum attached to an axle.

The system is usually filled with a glycol-ether based brake fluid (other fluids may also be used).

At one time, passenger vehicles commonly employed drum brakes on all four wheels. Later, disc brakes were used for the front and drum brakes for the rear. However disc brakes have shown better heat dissipation and greater resistance to 'fading' and are therefore generally safer than drum brakes. So four-wheel disc brakes hav become increasingly popular, replacing drums on all but the most basic vehicles. Many two-wheel vehicle designs, however, continue to employ a drum brake for the rear wheel.

The following description uses the terminology for and configuration of a simple disc brake.

System Operation

In a hydraulic brake system, when the brake pedal is pressed, a pushrod exerts force on the piston(s) in the master cylinder, causing fluid from the brake fluid reservoir to flow into a pressure chamber through a compensating port. This results in an increase in the pressure of the entire hydraulic system. This forces fluid through the hydraulic lines toward one or more calipers where it acts upon one or two caliper pistons sealed by one or more seated O-rings which prevent the escape of any fluid from around the piston.

The brake caliper pistons then apply force to the brake pads, which pushes them against the spinning rotor, and the friction between the pads and the rotor causes a braking torque to be generated, slowing the vehicle. Heat generated by this friction is either dissipated through vents and channels in the rotor or conducted through the pads, which are made of specialised heat-tolerant materials such as kevlar or sintered glass.

Subsequent release of the brake pedal/lever allows spring(s) to return the master piston(s) back into position. This relieves the hydraulic pressure on the caliper, allowing the brake piston in the caliper assembly to slide back into its housing and the brake pads to release the rotor.

The hydraulic braking system is designed as a closed system: unless there is a leak in the system, none of the brake fluid enters or leaves it, nor does the fluid get consumed through use.

An Example of a Hydraulic Brake System

Hydraulic brakes transfer energy to stop an object, normally a rotating axle. In a very simple brake system, with just two cylinders and a disc brake, the cylinders could be connected via tubes, with a piston inside the cylinders. The cylinders and tubes are filled with incompressible oil. The two cylinders have the same volume, but different diameters, and thus different cross-section areas. The one with the smallest diameter is called the master cylinder. The spinning disc brake will be placed down at the piston with the larger cross-section. Suppose the diameter of the master cylinder is half the diameter of the slave cylinder, so the master cylinder has a cross-section four times smaller. Now, if the piston in the master cylinder is pushed down 40 mm, with 10 newtons (N) of force, the slave piston will then move 10 mm, with a force of 40 N.

This force can be further increased by inserting a lever connected between the master piston, a pedal, and a pivot point. If the distance from the pedal to the pivot is three times the distance from the pivot to the connected piston, then it multiplies the pedal force by a factor of 3, when pushing down on the pedal. Now, if we push the pedal 120 mm down, with 10 N of force, a force of 30 N will then be applied to the master piston, and the slave piston will move the brake pad by 10 mm, with a force of 120 N.

Component Specifics

The brake pedal is a simple lever. One end is attached to the framework of the vehicle, a pushrod extends from a point along its length, and the foot pad is at the other end of the lever. The rod either extends to the master cylinder (manual brakes) or to the vacuum booster (power brakes).

In a four-wheel car, the FMVSS Standard 105, 1976; requires the master cylinder is divided internally into two sections, each of which pressurizes a separate hydraulic circuit. Each section supplies pressure to one circuit. Passenger vehicles typically have either a front/rear split brake system or a diagonal split brake system (the master cylinder in a motorcycle or scooter may only pressurize a single unit, which will be the front brake).

A front/rear split system uses one master cylinder section to pressurize the front caliper pistons and the other section to pressurize the rear caliper pistons. A split circuit braking system is now required by law in most countries for safety reasons; if one circuit fails, the other circuit can stop the vehicle.

Diagonal split systems were used initially on American Motors automobiles in the 1967 production year. The right front and left rear are served by one actuating piston while the left front and the right rear are served, exclusively, by a second actuating piston (both pistons pressurize their respective coupled lines from a single foot pedal). If either circuit fails, the other, with at least one front wheel braking (the front brakes provide most of the speed reduction) remains intact to stop the mechanically-damaged vehicle. Just before 1970, diagonally split systems had become universal for automobiles sold in the United States. This system was developed with front wheel drive cars suspension design to maintain better control and stability during a system failure.

The diameter and length of the master cylinder has a significant effect on the performance of the brake system. A larger diameter master cylinder delivers more hydraulic fluid to the caliper pistons, yet requires more brake pedal force and less brake pedal stroke to achieve a given deceleration. A smaller diameter master cylinder has the opposite effect. A master cylinder may also use differing diameters between the two sections to allow for increased fluid volume to one set of caliper pistons or the other.

Power Brakes

The vacuum booster or vacuum servo is used in most modern hydraulic brake systems which contain four wheels. The vacuum booster is attached between the master cylinder and the brake pedal and multiplies the braking force applied by the driver. These units consist of a hollow housing with a movable rubber diaphragm across the centre, creating two chambers. When attached to the low-pressure portion of the throttle body or intake manifold of the engine, the pressure in both chambers of the unit is lowered. The equilibrium created by the low pressure in both chambers keeps the diaphragm from moving until the brake pedal is depressed. A return spring keeps the diaphragm in the starting position until the brake pedal is applied. When the brake pedal is applied, the movement opens an air valve which lets in atmospheric pressure air to one chamber of the booster. Since the pressure becomes higher in one chamber, the diaphragm moves toward the lower pressure chamber with a force created by the area of the diaphragm and the differential pressure.

This force, in addition to the driver's foot force, pushes on the master cylinder piston. A relatively small diameter booster unit is required; for a very conservative 50% manifold vacuum, an assisting force of about 1500 N (200n) is produced by a 20 cm diaphragm with an area of 0.03 square metres. The diaphragm will stop moving when the forces on both sides of the chamber reach equilibrium. This can be caused by either the air valve closing (due to the pedal apply stopping) or if "run out" is reached. Run out occurs when the pressure in one chamber reaches atmospheric pressure and no additional force can be generated by the now stagnant differential pressure. After the run out point is reached, only the driver's foot force can be used to further apply the master cylinder piston.

The fluid pressure from the master cylinder travels through a pair of steel brake tubes to a pressure differential valve, sometimes referred to as a "brake failure valve", which performs two functions: it equalizes pressure between the two systems, and it provides a warning if one system loses pressure. The pressure differential valve has two chambers (to which the hydraulic lines attach) with a piston between them. When the pressure in either line is balanced, the piston does not move. If the pressure on one side is lost, the pressure from the other side moves the piston. When the piston makes contact with a simple electrical probe in the centre of the unit, a circuit is completed, and the operator is warned of a failure in the brake system.

From the pressure differential valve, brake tubing carries the pressure to the brake units at the wheels. Since the wheels do not maintain a fixed relation to the automobile, it is necessary to use hydraulic brake hose from the end of the steel line at the vehicle frame to the caliper at the wheel. Allowing steel brake tubing to flex invites metal fatigue and, ultimately, brake failure. A common upgrade is to replace the standard rubber hoses with a set which are externally reinforced with braided stainless-steel wires; these have negligible expansion under pressure and can give a firmer feel to the brake pedal with less pedal travel for a given braking effort.

Special Considerations

Air brake systems are bulky, and require air compressors and reservoir tanks. Hydraulic systems are smaller and less expensive.

Hydraulic fluid must be non-compressible. Unlike air brakes, where a valve is opened and air flows into the lines and brake chambers until the pressure rises sufficiently, hydraulic systems rely on a single stroke of a piston to force fluid through the system. If any vapor is introduced into the system it will compress, and the pressure may not rise sufficiently to actuate the brakes. Hydraulic braking systems are sometimes subjected to high temperatures during operation, such as when descending steep grades. For this reason, hydraulic fluid must resist vaporization at high temperatures.

Water vaporizes easily with heat and can corrode the metal parts of the system. If it gets into the brake lines, it can degrade brake performance dramatically. This is why light oils are often used as hydraulic fluids. Oil displaces water, protects plastic parts against corrosion, and can tolerate much higher temperatures before vaporizing.

"Brake fade" is a condition caused by overheating in which braking effectiveness reduces, and may be lost. It may occur for many reasons. The pads which engage the rotating part may become overheated and "glaze over", becoming so smooth and hard that they cannot grip sufficiently to slow the vehicle. Also, vaporization of the hydraulic fluid under temperature extremes or thermal distortion may cause the linings to change their shape and engage less surface area of the rotating part. Thermal distortion may also cause permanent changes in the shape of the metal components, resulting in a reduction in braking capability that requires replacement of the affected parts.

High-Density Solids Pump

High-density solids pumps are hydrostatically operating machines which displace the medium being pumped and thus create a flow.

High-Density Solids and Their Transport

High-density solids are mixtures of liquid and solid constituents. Each have their own typical physical and chemical characteristics such as specific weight, solids content, maximum particle size as well as how they behave, for example thixotropical, abrasively or adhesively. Typical examples for high-density solids are sludges and slurries.

Within industrial plants many substances with high proportion of solids - such as mechanically dewatered sewage sludges, filter cakes, bituminous coal sludges, waste and process sludges - have to be transported over relatively great distances to landfills or incineration plants.

In principle, long conveying distances can be bridged with mechanical conveying equipment or hydraulical high-density solids pumps.

Types of mechanical conveyors are belt, screw (worm), or scraper. They are suitable for nearly all types of high-density solids.

Hydraulical high-density solids pumps move sludges and slurries within an enclosed pipeline. However not every sludge or slurry is pumpable, depending on:

- Ratio: The mixture of solid and liquid components must be such that it results in a plastically deformable mass.
- Saturation: The volume of pore space in the sludges must be filled with sufficient quantity of solid matter for particle to be able to rest on particle via a plastically viscous liquid and for interstices to be filled.
- Gas component: As a consequence of pressure fluctuations knocking in the pipe may happen. By injection of about 2% air, sludges which do not naturally have a gas component can be brought to a better consistency.

Types of High-Density Solids Pumps

Depending on how the displacement principle is implemented, a distinction can be drawn between rotary and reciprocating pumps. The rotary circulation pumps includes eccentric screw pumps, centrifugal pumps, and squeezed tube (peristaltic) pumps. Reciprocating pumps include plunger, diaphragm, and piston pumps.

The high-density solids pumps which can be used in the widest range of situations is the hydraulically powered piston (or reciprocating) pump. It can be realised as single-cylinder or two-cylinder pump. In the latter case, power is transferred to the material via delivery pistons working in push-pull-mode. While one delivery piston sucks material from the feed hopper into the cylinder, the second piston simultaneously pushes the material in the other delivery cylinder and into the delivery line.

Type: Piston Pump with Transfer Tube (Two-Cylinder Pump)

Construction Details: The most characteristic element of the transfer tube pump is the so-called "S" transfer tube installed inside the feed hopper. The job of the transfer tube is to control the flow of high-density solids - in other words, the pump cylinder which is currently pushing the material is connected to the delivery line by such tube. At (or nearby) the end of each piston stroke the S-shaped tube which is connected to the delivery line is swung over to the other delivery cylinder with the aid of two switch cylinders. A hydraulic circuit is used for synchronizing the position of the transfer tube and the movements of the two delivery pistons.

The S tube should swing over rapidly and not suffer undue wear. This is effected with two large-sized plunger cylinders and a so-called spectacle plate (so named from its shape) including seal rings.

In cross section the S-tube is circular with a diameter which tapers in the direction of flow. This keeps the risk of clogging down. The sealing faces of the spectacle plate and of the seal ring lie parallel with the swivelling movement of the S-tube, so that they cannot be damaged when foreign bodies are cut through or get stuck.

This pump does not need check valves.

Operating Features

Piston pumps with S-tube are suitable for concrete and mortar, for sludges with a solids content of up to 50% by weight and a fluctuating particle size distribution but also for conveying fly ash, coal or minerals in a suspension.

The large cross section of the holes in the spectacle plate means that with a delivery cylinder diameter of e.g. 200 mm (7.874 in) it is possible to pump high-density solids with a mean grain size of as much as 80 mm (3.150 in). The maximum diameter of individual foreign bodies may be as high as 60% of the diameter of the delivery line - in this example this would be 120 mm (4.724 in).

Type: Piston Pump with Seat Valves (Two-Cylinder Pump)

Construction details: The steel housing of the seat valve pump contains four hydraulically controlled seat valves, two suction and two delivery valves for the two hydraulic cylinders. The suction and discharge valves are synchronized with the hydraulics of the delivery pistons - this ensures that the contents of the delivery cylinder is equal at all times to the volume being pumped. Once the "sucking" piston reaches its end position, the corresponding suction or delivery valve is simultaneously closed or opened. If there is overpressure in the delivery line, the suction valve will close first. This prevents the pumped medium from being pushed out of the pressure line back into the hopper.

This pump does not need check valves.

Operating Features

Seat valve pumps are suitable for an especially even pumping of media with a solids content of up to 50% and for the high-pressure pumping of paste-like industrial media such as sludges. Since valves are used in this two-cylinder pump it is suitable for grain sizes up to 8mm at most.

Which type of valve is selected will depend on the high-density solids to be conveyed. For materials containing grains with a solids content up to 50% a sharp-edged metal valve seat is required. With low-viscosity, rather watery fine sludges the choice would be a large-area elastomeric valve seal.

In one hydraulic control variant the valve is opened passively by the thrust of the medium. In this case the discharge valve has a non-return function which prevents flowback from the pressure line. At the same time the high-density solids are precompacted to close to the line pressure before the discharge valve opens. This means that pipe knocking resulting from pressure pulsations can be avoided.

Type: Ball-Valve Pump (Two-Cylinder Pump)

Construction Details: During the suction stroke of the delivery piston the medium is sucked in via the open suction ball valve. At the same time the ball valve on the delivery line is pushed into the valve seating by the underpressure resulting from the suction (automatic valves). In parallel with this, the second delivery piston is making the pressure stroke and forces the medium through the discharge ball valve and into the delivery line. The suction ball valve is pushed into its seating by the delivery line pressure and thereby closes off the connection to pump intake.

Operating Features

The two-cylinder ball valve pump is suitable for in the low to middle pressure range for pumping highly liquid to paste-like media provided these can be sucked in through the valve openings - for example mortars, mineral and sewage sludges.

The ball valve pump is to all intents and purposes insensitive to corrosive and abrasive media since the ball valves do not have to be opened or closed from the outside.

Type: Single-cylinder piston pump

Construction Details: With a single-cylinder piston pump the medium is "stuffed" out of a feed chute into the delivery cylinder and by the cylinder into the delivery line on the forward stroke. Depending on the level of the delivery pressure in the pipe and on the flow properties of the medium, the delivery flange of the pump will be fitted with a flat slide valve or a lamellar non-return valve in order to prevent flowback of the material on the return stroke. Which delivery piston is actually selected will depend on the material which is to be pumped: Free flowing media call for perturbane sealing elements while bulky materials which have to be chopped up during pumping mean that hardened cutting edges are required.

Operating Features

The single-cylinder pump is suitable for bulk materials with an awkward coarse composition - for example wood chips, organic waste, hazardous waste, or shredded paper. If such unsaturated material is to be conveyed over large distances, one possible approach is the hybrid feed system. Here compressed air is injected into the delivery line. It presses the plug of material into the pipe. As the compressed air expands the material being conveyed is loosened and separated. As distance increases from the air injection point, what was initially a series of plugs gradually turns into a continuous airborne stream. The amount of air required and the appropriate pipe diameter depend on the type and quantity of matter being conveyed.

Parts of High-Density Solids Piston Pumps

Main parts are:

- feed hopper
- drive cylinder, delivery cylinder and hydraulic system
- seals
- electrical and control equipment

  and - depending on the type:
- transfer tube

- valves

as well as peripheral components such as:

- bunker (with / without sliding frames)
- screw conveyors, pre-press equipment
- pipework
- air injection
- boundary layer injection
- systems for pressure (pulsation) damping

Solids Pump Selection

The special kind of medium "high-density solids" and the wide range of applications require different construction principles and hence pump types. Following table permits a pre-selection depending on the media characteristics:

Type	*Medium (High-Density Solids)*			
	Watery	*Stiff*	*Bulky*	*with Foreign Bodies*
Transfer type pump	+	+ +	-	+
Seat valve pump	+ +	+	-	-
Ball valve pump	+	o	-	-
Single-cylinder pump	-	+ +	+ +	+ +

+ + very suitable, + suitable, o with reservations, - not suitable

Applications

The hydraulic pumping of solids with high solids contents offers a range of possibilities in process engineering. Oil-hydraulically powered two-cylinder piston pumps with S-transfer tube have already been used for many years in civil engineering for transporting concrete. This technology has been transferred to further areas of application in various sectors of industry.

Sewage Treatment Plants

Mechanically dewatered sewage sludge can be pumped directly into furnaces via a pipe network. This avoids direct contact with the sludges and prevents from contamination and odour emission.

Waste Management (Hazardous Waste)

Hazardous and highly flammable substances are processed physically, chemically and biologically in order to reduce the waste materials. With the advantage of pipeline conveyance, high density solids pumps commonly transport such material.

Mining

Real tunnelling (in other words, not cut-and-cover) has increased palpably in recent years. Conveying mine water, soil and tailings to the tunnel entrance is a need.

Boundary Conditions

Transport Routes: Various aspects have to be considered in the planning and processing of conveying installations. This includes the space requirements, the accessibility in case of maintenance or cleaning purposes, as well as horizontal or vertical transport route for material discharge.

Environmental Protection (Noise and Odour Reduction)

Due to the nature of the material and the conveying method, odour nuisances is to be expected. Choosing the appropriate conveying method can reduce the environmental pollution.

Energy Expenditure

Three forces have to be overcome when conveying with high density solids pumps:

1. The conveying distance (cf. Record held by Concrete pump)
2. Friction between material and pipeline, especially in bends and reduction joints.
3. Inner frictional force caused by deformation of the material.

Calculation example:

- Conveyance suspension
- Delivery line of 200 mm (8 in) diameter (DN200)
- High-density solids with mineral or carbon content as a mixture from coarse and fine grained material.
- max. particle size 50 mm (1.969 in)
- Dry solids content up to 80 percent by weight

Energy consumption = approx. 50 Wh (180 kJ) per tonne of conveyed material and per kilometre (280 Btu per long ton or 250 Btu per short ton per mile).

7

Dimensional Analysis

In physics and all science, dimensional analysis is the practice of checking relations among physical quantities by identifying their dimensions. The dimension of any physical quantity is the combination of the basic physical dimensions that compose it. Some fundamental physical dimensions are length, mass, time, and electric charge. Other physical quantities can be expressed in terms of these fundamental physical dimensions. For example, speed has the dimension length (or distance) per unit of time, and may be measured in metres per second, miles per hour, or other units. Similarly electrical current is electrical charge per unit time (flow rate of charge) and is measured in coulombs (a unit of electrical charge) per second, or equivalently, amperes. Dimensional analysis is based on the fact that a physical law must be independent of the units used to measure the physical variables. A straightforward practical consequence is that any meaningful equation (and any inequality and inequation) must have the same dimensions on the left and right sides. Checking this is the basic way of performing dimensional analysis.

Dimensional analysis is routinely used to check the plausibility of derived equations and computations. It is also used to form reasonable hypotheses about complex physical situations that can be tested by experiment or by more developed theories of the phenomena, and to categorize types of physical quantities and units based on their relations to or dependence on other units, or their dimensions if any.

Great Principle of Similitude

The basic principle of dimensional analysis was known to Isaac Newton (1686) who referred to it as the "*Great Principle of Similitude*".

James Clerk Maxwell played a major role in establishing modern use of dimensional analysis by distinguishing mass, length, and time as fundamental units, while referring to other units as derived. The 19th-century French mathematician Joseph Fourier made important contributions based on the idea that physical laws like $F = ma$ should be independent of the units employed to measure the physical variables. This led to the conclusion that meaningful laws must be homogeneous equations in their various units of measurement, a result which was eventually formalized in the Buckingham ð theorem. This theorem describes how every physically meaningful equation involving n variables can be equivalently rewritten as an equation of $n - m$ dimensionless parameters, where m is the rank of the dimensional matrix. Furthermore, and most importantly, it provides a method for computing these dimensionless parameters from the given variables.

A dimensional equation can have the dimensions reduced or eliminated through nondimensionalization, which begins with dimensional analysis, and involves scaling quantities by characteristic units of a system or natural units of nature. This gives insight into the fundamental properties of the system, as illustrated in the examples below.

Definition

The dimension of a physical quantity can be expressed as a product of the basic physical dimensions mass, length, time, electric charge, and absolute temperature, represented by sans-serif symbols M, L, T, Q, and Θ, respectively, each raised to a rational power.

The SI standard recommends the usage of the following dimensions and corresponding symbols: mass (M), length (L), time (T), electrical current (I), absolute temperature (Θ), amount of substance (N) and luminous intensity (J).

The term *dimension* is more abstract than *scale* unit: *mass* is a dimension, while kilograms are a scale unit (choice of standard) in the mass dimension.

As examples, the dimension of the physical quantity speed is *length/time* (L/T or LT^{-1}), and the dimension of the physical quantity force is "mass × acceleration" or "mass×(length/time)/time" (ML/T^2 or MLT^{-2}). In principle, other dimensions of physical quantity could be defined as "fundamental" (such as momentum or energy or electric current) in lieu of some of those shown above. Most physicists do not recognise temperature, Θ, as a fundamental dimension of physical quantity since it essentially expresses the energy per particle per degree of freedom, which can be expressed in terms of energy (or mass, length, and time). Still others do not recognise electric current, I, as a separate fundamental dimension of physical quantity, since it has been expressed in terms of mass, length, and time in unit systems such as the cgs system. There are also physicists that have cast doubt on the very existence of incompatible fundamental dimensions of physical quantity.

The unit of a physical quantity and its dimension are related, but not identical concepts. The units of a physical quantity are defined by convention and related to some standard; e.g., length may have units of metres, feet, inches, miles or micrometres; but any length always has a dimension of L, independent of what units are arbitrarily chosen to measure it. Two different units of the same physical quantity have conversion factors that relate them. For example: 1 in = 2.54 cm; then (2.54 cm/in) is the conversion factor, and is itself dimensionless and equal to one. Therefore multiplying by that conversion factor does not change a quantity. Dimensional symbols do not have conversion factors.

Mathematical Properties

The dimensions that can be formed from a given collection of basic physical dimensions, such as M, L, and T, form an abelian group: The identity is written as 1; $L^0 = 1$, and the inverse to L is 1/L or L^{-1}. L raised to any rational power p is a member of the group, having an inverse of L^{-p} or $1/L^p$. The operation of the group is multiplication, having the usual rules for handling exponents ($L^n \times L^m = L^{n+m}$).

This group can be described as a vector space over the rational numbers, with for example dimensional symbol $M^iL^jT^k$ corresponding to the vector (i, j, k). When physical measured quantities (be they like-dimensioned or unlike-dimensioned) are multiplied or divided by one other, their dimensional units are likewise multiplied or divided; this corresponds to addition or subtraction in the vector space. When measurable quantities are raised to a rational power, the same is done to the dimensional symbols attached to those quantities; this corresponds to scalar multiplication in the vector space.

A basis for a given vector space of dimensional symbols is called a set of fundamental units or fundamental dimensions, and all other vectors are called derived units. As in any vector space, one may choose different bases, which yields different systems of units (e.g., choosing whether the unit for charge is derived from the unit for current, or vice versa).

The group identity 1, the dimension of dimensionless quantities, corresponds to the origin in this vector space.

The set of units of the physical quantities involved in a problem correspond to a set of vectors (or a matrix). The kernel describes some number (e.g., m) of ways in which these vectors can be combined to produce a zero vector. These correspond to producing (from the measurements) a number of dimensionless quantities, $\{\pi_1,...,\pi_m\}$. (In fact these ways completely span the null subspace of another different space, of powers of the measurements.) Every possible way of multiplying (and exponentiating) together the measured quantities to produce something with the same units as some derived quantity X can be expressed in the general form

$$X = \prod_{i=1}^{m} (\pi_i)^{k_i}.$$

Consequently, every possible commensurate equation for the physics of the system can be rewritten in the form

$$f(\pi_1, \pi_2, ..., \pi_m) = 0.$$

Knowing this restriction can be a powerful tool for obtaining new insight into the system.

Mechanics

In mechanics, the dimension of any physical quantity can be expressed in terms of the fundamental dimensions (or *base dimensions*) M, L, and T – these form a 3-dimensional vector space. This is not the only possible choice, but it is the one most commonly used. For example, one might choose force, length and mass as the base dimensions (as some have done), with associated dimensions F, L, M; this corresponds to a different basis, and one may convert between these representations by a change of basis.

The choice of the base set of dimensions is, thus, partly a convention, resulting in increased utility and familiarity. It is, however, important to note that the choice of the set of dimensions cannot be chosen arbitrarily – it is not *just* a convention – because the dimensions must form a basis: they must span the space, and be linearly independent.

For example, F, L, M form a set of fundamental dimensions because they form an equivalent basis to *M, L, T:* the former can be expressed as $[F=ML/T^2]$,L,M while the latter can be expressed as M,L,$[T=(ML/F)^{1/2}]$.

On the other hand, using length, velocity and time (*L, V, T*) as base dimensions will not work well (they do not form a set of fundamental dimensions), for two reasons:

- There is no way to obtain mass – or anything derived from it, such as force – without introducing another base dimension (thus these do not *span the space*).
- Velocity, being derived from length and time (V=L/T), is redundant (the set is not *linearly independent*).

Other Fields of Physics and Chemistry

Depending on the field of physics, it may be advantageous to choose one or another extended set of dimensional symbols. In electromagnetism, for example, it may be useful to use dimensions of M, L, T, and Q, where Q represents quantity of electric charge. In thermodynamics, the base set of dimensions is often extended to include a dimension for temperature, Θ. In chemistry the number of moles of substance (loosely, but not precisely, related to the number of molecules or atoms) is often involved and a dimension for this is used as well. In the interaction of relativistic plasma with strong laser pulses a dimensionless relativistic similarity parameter connected with the symmetry properties of the collisionless Vlasov equation is constructed from the plasma electron and critical densities in addition to the electromagnetic vector potential. The choice of the dimensions or even the number of dimensions to be used in different fields of physics is to some extent arbitrary, but consistency in use and ease of communications are very important.

Commensurability

The most basic rule of dimensional analysis is that of dimensional homogeneity, first formulated by Newton, who called it the "great principle of similitude":

Only *commensurable* quantities (quantities with the same dimensions) may be *compared, equated, added,* or *subtracted.*

However, as stated above, the dimensions form a multiplicative group; consequently:

One may take *ratios* of *incommensurable* quantities (quantities with different dimensions), and *multiply* or *divide* them.

For example, it makes no sense to ask if 1 hour is more, the same, or less than 1 kilometre, as these have different dimensions, nor to add 1 hour to 1 kilometre. On the other hand, if an object travels

100 km in 2 hours, one may divide these and conclude that the object's average speed was 50 km/h.

The rule implies that in a physically meaningful *expression* only quantities of the same dimension can be added, subtracted, or compared. For example, if m_{man}, m_{rat} and L_{man} denote, respectively, the mass of some man, the mass of a rat and the length of that man, the dimensionally homogeneous expression $m_{man} + m_{rat}$ is meaningful, but the heterogeneous expression $m_{man} + L_{man}$ is meaningless. However, m_{man}/L^2_{man} is fine. Thus, dimensional analysis may be used as a sanity check of physical equations: the two sides of any equation must be commensurable or have the same dimensions.

Even when two physical quantities have identical dimensions, it may nevertheless be meaningless to compare or add them. For example, although torque and energy share the dimension ML^2/T^2, they are fundamentally different physical quantities. To compare, add, or subtract quantities with the same dimensions but expressed in different units, the standard procedure is to first convert them all to the same units. For example, to compare 32 metres with 35 yards, use 1 yard = 0.9144 m to convert 35 yards to 32.004 m.

Polynomials and Transcendental Functions

Scalar arguments to transcendental functions such as exponential, trigonometric and logarithmic functions, or to inhomogeneous polynomials, must be dimensionless quantities. (Note: this requirement is somewhat relaxed in Siano's orientational analysis described below, in which the square of certain dimensioned quantities are dimensionless.) While most mathematical identities about dimensionless numbers translate in a straightforward manner to dimensional quantities, care must be taken with logarithms of ratios: the identity log(a/b) = log a " log b, where the logarithm is taken in any base, holds for dimensionless numbers a and b, but it does *not* hold if a and b are dimensional, because in this case the left-hand side is well-defined but the right-hand side is not.

Similarly, while one can evaluate monomials (x^n) of dimensional quantities, one cannot evaluate polynomials of mixed degree with dimensionless coefficients on dimensional quantities: for x^2, the expression $(3\text{ m})^2 = 9\text{ m}^2$ makes sense (as an area), while for $x^2 + x$, the expression $(3\text{ m})^2 + 3\text{ m} = 9\text{ m}^2 + 3\text{ m}$ does not make sense.

However, polynomials of mixed degree can make sense if the coefficients are suitably chosen physical quantities that are not dimensionless. For example,

$$\frac{1}{2}\cdot\left(-32\frac{\text{foot}}{\text{second}^2}\right)\cdot t^2+\left(500\frac{\text{foot}}{\text{second}}\right)\cdot t.$$

This is the height to which an object rises in time t if the acceleration of gravity is 32 feet per second per second and the initial upward speed is 500 feet per second. It is not even necessary for t to be in *seconds*. For example, suppose $t = 0.01$ minutes. Then the first term would be

$$\frac{1}{2}\cdot\left(-32\frac{\text{foot}}{\text{second}^2}\right)\cdot(0.01\text{ minute})^2$$

$$=\frac{1}{2}\cdot-32\cdot(0.01^2)\left(\frac{\text{minute}}{\text{second}}\right)^2\cdot\text{foot}$$

$$=\frac{1}{2}\cdot-32\cdot(0.01^2)\cdot 60^2\cdot\text{foot}.$$

Incorporating Units

The value of a dimensional physical quantity Z is written as the product of a unit $[Z]$ within the dimension and a dimensionless numerical factor, n.

$$Z = n\times[Z] = n[Z]$$

In a strict sense, when like-dimensioned quantities are added or subtracted or compared, these dimensioned quantities must be expressed in consistent units so that the numerical values of these quantities may be directly added or subtracted. But, in concept, there is no problem adding quantities of the same dimension expressed in different units. For example, 1 metre added to 1 foot *is* a length, but it would not be correct to add 1 to 1 to get the result. A conversion factor, which is a ratio of like-dimensioned quantities and is equal to the dimensionless unity, is needed:

$$1\text{ ft} = 0.3048\text{ m} \text{ is identical to } 1=\frac{0.3048\text{ m}}{1\text{ ft}}.$$

The factor $0.3048\frac{\text{m}}{\text{ft}}$ is identical to the dimensionless 1, so multiplying by this conversion factor changes nothing. Then when adding two quantities of like dimension, but expressed in different units, the appropriate conversion factor, which is essentially the dimensionless 1, is used to convert the quantities to identical units so that their numerical values can be added or subtracted. Only in this manner is it meaningful to speak of adding like-dimensioned quantities of differing units.

Position vs Displacement

Some discussions of dimensional analysis implicitly describe all quantities as mathematical vectors. (In mathematics scalars are considered a special case of vectors; the emphasis here is that vectors are closed under addition, subtraction, and scalar multiplication, and permit scalar division.). This assumes an implicit point of reference – an origin. While this is useful and often perfectly adequate, allowing many important errors to be caught, it can fail to model certain aspects of physics. A more rigorous approach requires distinguishing between position and displacement (or moment in time versus duration, or absolute temperature versus temperature change).

Consider points on a line, each with a position with respect to a given origin, and distances among them. Positions and displacements all have units of length, but their meaning is not interchangeable:

- adding two displacements should yield a new displacement (walking ten paces then twenty paces gets you thirty paces forward),
- adding a displacement to a position should yield a new position (walking one block down the street from an intersection gets you to the next intersection),
- subtracting two positions should yield a displacement,
- but one may *not* add two positions.

This illustrates the subtle distinction between *affine* quantities (ones modelled by an affine space, such as position) and *vector* quantities (ones modelled by a vector space, such as displacement).

- Vector quantities may be added to each other, yielding a new vector quantity, and a vector quantity may be added to a suitable affine quantity (a vector space *acts on* an affine space), yielding a new affine quantity.
- Affine quantities cannot be added, but may be subtracted, yielding *relative* quantities which are vectors, and these *relative differences* may then be added to each other or to an affine quantity.

Properly then, positions have dimension of *affine* length, while displacements have dimension of *vector* length. To assign a number to an *affine* unit, one must not only choose a unit of measurement, but also a point of reference, while to assign a number to a *vector* unit only requires a unit of measurement.

Thus some physical quantities are better modelled by vectorial quantities while others tend to require affine representation, and the distinction is reflected in their dimensional analysis.

This distinction is particularly important in the case of temperature, for which the numeric value of absolute zero is not the origin 0 in some scales. For absolute zero,

0 K = –273.15 °C = –459.67 °F = 0 °R,

but for temperature differences,

1 K = 1 °C ≠ 1 °F = 1 °R.

(Here °R refers to the Rankine scale, not the Réaumur scale). Unit conversion for temperature differences is simply a matter of multiplying by, e.g., 1 °F / 1 K. But because some of these scales have origins that do not correspond to absolute zero, conversion from one temperature scale to another requires accounting for that. As a result, simple dimensional analysis can lead to errors if it is ambiguous whether 1 K means the absolute temperature equal to –272.15 °C, or the temperature difference equal to 1 °C.

Orientation and Frame of Reference

Similar to the issue of a point of reference is the issue of orientation: a displacement in 2 or 3 dimensions is not just a length, but is a length together with a *direction*. (This issue does not arise in 1 dimension, or rather is equivalent to the distinction between positive and negative.) Thus, to compare or combine two dimensional quantities in a multi-dimensional space, one also needs an orientation: they need to be compared to a frame of reference.

This leads to the extensions discussed below, namely Huntley's directed dimensions and Siano's orientational analysis.

Other Uses

Dimensional analysis is also used to derive relationships between the physical quantities that are involved in a particular phenomenon that one wishes to understand and characterize. It was used for the first time (Pesic 2005) in this way in 1872 by Lord Rayleigh, who was trying to understand why the sky is blue. Rayleigh first published the technique is his book "theory of sound" from 1877.

Examples

A simple example: period of a harmonic oscillator:

What is the period of oscillation T of a mass m attached to an ideal linear spring with spring constant k suspended in gravity of strength g? That period is the solution for T of some dimensionless equation in the variables T, m, k, and g. The four quantities have the following dimensions: T [T]; m [M]; k [M/T²]; and g [L/T²]. From these we can form only one dimensionless product of powers of our chosen variables, $G_1 = T^2 k / m$ [$T^2 \cdot M/T^2 / M = 1$], and putting $G_1 = C$ for some dimensionless constant C gives the dimensionless equation sought. The dimensionless product of powers of variables is sometimes referred to as a dimensionless group of variables; here the term "group" means "collection" rather than mathematical group. They are often called dimensionless numbers as well.

Note that the variable g does not occur in the group. It is easy to see that it is impossible to form a dimensionless product of powers that combines g with k, m, and T, because g is the only quantity that involves the dimension L. This implies that in this problem the g is irrelevant. Dimensional analysis can sometimes yield strong statements about the *irrelevance* of some quantities in a problem, or the need for additional parameters. If we have chosen enough variables to properly describe the problem, then from this argument we can conclude that the period of the mass on the spring is independent of g: it is the same on the earth or the moon. The equation demonstrating the existence of a product of powers for our problem can be written in an entirely equivalent way: $T = \kappa\sqrt{m / k}$, for some dimensionless constant κ (equal to $\sqrt{C}$ from the original dimensionless equation).

When faced with a case where dimensional analysis rejects a variable (g, here) that one intuitively expects to belong in a physical description of the situation, another possibility is that the rejected variable is in fact relevant, but that some other relevant variable has been omitted, which might combine with the rejected variable to form a dimensionless quantity. That is, however, not the case here.

When dimensional analysis yields only one dimensionless group, as here, there are no unknown functions, and the solution is said to be "complete" – although it still may involve unknown dimensionless constants, such as κ.

A More Complex Example: Energy of a Vibrating Wire

Consider the case of a vibrating wire of length ℓ (L) vibrating with an amplitude A (L). The wire has a linear density ρ (M/L) and is under tension s (ML/T^2), and we want to know the energy E (ML^2/T^2) in the wire. Let π_1 and π_2 be two dimensionless products of powers of the variables chosen, given by

$$\pi_1 = E / As$$
$$\pi_2 = \ell / A.$$

The linear density of the wire is not involved. The two groups found can be combined into an equivalent form as an equation

$$F(E / As, \ell / A) = 0,$$

where F is some unknown function, or, equivalently as

$$E = Asf(\ell / A),$$

where f is some other unknown function. Here the unknown function implies that our solution is now incomplete, but dimensional analysis has given us something that may not have been obvious: the energy is proportional to the first power of the tension. Barring further analytical analysis, we might proceed to experiments to discover the form for the unknown function f. But our experiments are simpler than in the absence of dimensional analysis. We'd perform none to verify that the energy is proportional to the tension. Or perhaps we might guess that the energy is proportional to ℓ, and so infer that $E = \ell s$. The power of dimensional analysis as an aid to experiment and forming hypotheses becomes evident.

The power of dimensional analysis really becomes apparent when it is applied to situations, unlike those given above, that are more complicated, the set of variables involved are not apparent, and the underlying equations hopelessly complex. Consider, for example, a small pebble sitting on the bed of a river. If the river flows fast enough, it will actually raise the pebble and cause it to flow along with the water. At what critical velocity will this occur? Sorting out the guessed variables is not so easy as before. But dimensional analysis can be a powerful aid in understanding problems like this, and is usually the very first tool to be applied to complex problems where the underlying equations and constraints are poorly understood. In such cases, the answer may depend on a dimensionless number such as the Reynolds number, which may be interpreted by dimensional analysis.

Extensions

Huntley's Extension: Directed Dimensions: Huntley (Huntley 1967) has pointed out that it is sometimes productive to refine our concept of dimension. Two possible refinements are:

- The magnitude of the components of a vector are to be considered dimensionally distinct. For example, rather than an undifferentiated length unit L, we may have L_x represent length in the x direction, and so forth. This requirement stems ultimately from the requirement that each component of a physically meaningful equation (scalar, vector, or tensor) must be dimensionally consistent.

- Mass as a measure of quantity is to be considered dimensionally distinct from mass as a measure of inertia.

As an example of the usefulness of the first refinement, suppose we wish to calculate the distance a cannonball travels when fired with a vertical velocity component V_y and a horizontal velocity component V_x, assuming it is fired on a flat surface. Assuming no use of directed lengths, the quantities of interest are then V_x, V_y, both dimensioned as L/T, R, the distance travelled, having dimension L, and g the downward acceleration of gravity, with dimension L/T^2

With these four quantities, we may conclude that the equation for the range R may be written:

$$R \propto V_x^a V_y^b g^c.$$

Or dimensionally

$$L = (L/T)^{a+b}(L/T^2)^c$$

from which we may deduce that $a+b+c=1$ and $a+b+2c=0$, which leaves one exponent undetermined. This is to be expected since we have two fundamental quantities L and T and four parameters, with one equation.

If, however, we use directed length dimensions, then V_x will be dimensioned as L_x/T, V_y as L_y/T, R as L_x and g as L_y/T^2. The dimensional equation becomes:

$$L_x = (L_x/T)^a(L_y/T)^b(L_y/T^2)^c$$

and we may solve completely as $a=1, b=1$ and $c=-1$. The increase in deductive power gained by the use of directed length dimensions is apparent.

In a similar manner, it is sometimes found useful (e.g., in fluid mechanics and thermodynamics) to distinguish between mass as a measure of inertia (inertial mass), and mass as a measure of quantity (substantial mass). For example, consider the derivation of Poiseuille's Law. We wish to find the rate of mass flow of a viscous fluid through a circular pipe. Without drawing distinctions between inertial and substantial mass we may choose as the relevant variables

- $\dot{m}$ the mass flow rate with dimensions M / T
- p_x the pressure gradient along the pipe with dimensions M / L^2T^2
- ρ the density with dimensions M / L^3
- η the dynamic fluid viscosity with dimensions M / LT
- r the radius of the pipe with dimensions L

There are three fundamental variables so the above five equations will yield two dimensionless variables which we may take to be $\pi_1 = \dot{m} / \eta r$ and $\pi_2 = p_x \rho r^5 / \dot{m}^2$ and we may express the dimensional equation as

$$C = \pi_1 \pi_2^a = \left(\frac{\dot{m}}{\eta r}\right)\left(\frac{p_x \rho r^5}{\dot{m}^2}\right)^a$$

where C and a are undetermined constants. If we draw a distinction between inertial mass with dimensions M_i and substantial mass with dimensions M_s, then mass flow rate and density will use substantial mass as the mass parameter, while the pressure gradient and coefficient of viscosity will use inertial mass. We now have four fundamental parameters, and one dimensionless constant, so that the dimensional equation may be written:

$$C = \frac{p_x \rho r^4}{\eta \dot{m}}$$

where now only C is an undetermined constant (found to be equal to $\pi / 8$ by methods outside of dimensional analysis). This equation may be solved for the mass flow rate to yield Poiseuille's law.

Siano's Extension: Orientational Analysis

Huntley's extension has some serious drawbacks:

- It does not deal well with vector equations involving the *cross product,*
- nor does it handle well the use of *angles* as physical variables.

It also is often quite difficult to assign the L, L_x, L_y, L_z, symbols to the physical variables involved in the problem of interest. He invokes a procedure that involves the "symmetry" of the physical problem. This is often very difficult to apply reliably: It is unclear as to what parts of the problem that the notion of "symmetry" is being invoked. Is it the symmetry of the physical body that forces are acting upon, or to the points, lines or areas at which forces are being applied?

What if more than one body is involved with different symmetries? Consider the spherical bubble attached to a cylindrical tube, where one wants the flow rate of air as a function of the pressure difference in the two parts. What are the Huntley extended dimensions of the viscosity of the air contained in the connected parts? What are the extended dimensions of the pressure of the two parts? Are they the same or different? These difficulties are responsible for the limited application of Huntley's addition to real problems.

Angles are, by convention, considered to be dimensionless variables, and so the use of angles as physical variables in dimensional analysis can give less meaningful results. As an example, consider the projectile problem mentioned above. Suppose that, instead of the x- and y-components of the initial velocity, we had chosen the magnitude of the velocity v and the angle θ at which the projectile was fired. The angle is, by convention, considered to be dimensionless, and the magnitude of a vector has no directional quality, so that no dimensionless variable can be composed of the four variables g, v, R, and θ. Conventional analysis will correctly give the powers of g and v, but will give no information concerning the dimensionless angle θ.

Siano (1985-I, 1985-II) has suggested that the directed dimensions of Huntley be replaced by using *orientational symbols* $1_x\ 1_y\ 1_z$ to denote vector directions, and an orientationless symbol 1_0. Thus, Huntley's 1_x becomes $L\ 1_x$ with L specifying the dimension of length, and 1_x specifying the orientation. Siano further shows that the orientational symbols have an algebra of their own. Along with the requirement that $1_i^{-1} = 1_i$, the following multiplication table for the orientation symbols results:

	1_0	1_x	1_y	1_z
1_0	1_0	1_x	1_y	1_z
1_x	1_x	1_0	1_z	1_y
1_y	1_y	1_z	1_0	1_x
1_z	1_z	1_y	1_x	1_0

Note that the orientational symbols form a group (the Klein four-group or "Viergruppe"). In this system, scalars always have the same orientation as the identity element, independent of the "symmetry of the problem." Physical quantities that are vectors have the orientation expected: a force or a velocity in the z-direction has the orientation of 1_z. For angles, consider an angle θ that lies in the z-plane. Form a right triangle in the z plane with θ being one of the acute angles. The side of the right triangle adjacent to the angle then has an orientation 1_x and the side opposite has an orientation 1_y. Then, since $\tan(\theta) = 1_y/1_x = \theta + \ldots$ we conclude that an angle in the xy plane must have an orientation $1_y/1_x = 1_z$, which is not unreasonable. Analogous reasoning forces the conclusion that $\sin(\theta)$ has orientation 1_z while $\cos(\theta)$ has orientation 1_0. These are different, so one concludes (correctly), for example, that there are no solutions of physical equations that are of the form $a\cos(\theta)+b\sin(\theta)$, where a and b are real scalars. Note that an expression such as $\sin(\theta + \pi/2) = \cos(\theta)$ is not dimensionally inconsistent since it is a special case of the sum of angles formula and should properly be written:

$$\sin(a1_z + b1_z) = \sin(a1_z)\cos(b1_z) + \sin(b1_z)\cos(a1_z)$$

which for $a = \theta$ and $b = \pi/2$ yields $\sin(\theta 1_z + (\pi/2)1_z) = 1_z\cos(\theta 1_z)$. Physical quantities may be expressed as complex numbers (e.g. $e^{i\theta}$) which imply that the complex quantity i has an orientation equal to that of the angle it is associated with (1_z in the above example).

The assignment of orientational symbols to physical quantities and the requirement that physical equations be orientationally homogeneous can actually be used in a way that is similar to dimensional analysis to derive a little more information about acceptable solutions of physical problems. In this approach one sets up the dimensional equation and solves it as far as one can. If the lowest power of a physical variable is fractional, both sides of the solution is raised to a power such that all powers are integral. This puts it into “normal form”. The orientational equation is then solved to give a more restrictive condition on the unknown powers of the orientational symbols, arriving at a solution that is more complete than the one that dimensional analysis alone gives. Often the added information is that one of the powers of a certain variable is even or odd.

As an example, for the projectile problem, using orientational symbols, θ, being in the *xy*-plane will thus have dimension 1_z and the range of the projectile R will be of the form:

$$R = g^a v^b \theta^c \text{ which means } L1_x \sim \left(\frac{L1_y}{T^2}\right)^a \left(\frac{L}{T}\right)^b 1_z^c.$$

Dimensional homogeneity will now correctly yield $a = -1$ and $b = 2$, and orientational homogeneity requires that c be an odd integer. In fact the required function of theta will be $\sin(\theta)\cos(\theta)$ which is a series of odd powers of θ.

It is seen that the Taylor series of sin(θ) and cos(θ) are orientationally homogeneous using the above multiplication table, while expressions like cos(θ) + sin(θ) and exp(θ) are not, and are (correctly) deemed unphysical.

It should be clear that the multiplication rule used for the orientational symbols is not the same as that for the cross product of two vectors. The cross product of two identical vectors is zero, while the product of two identical orientational symbols is the identity element.

Percentages and Derivatives

Percentages are dimensionless quantities, since they are ratios of two quantities with the same dimensions.

Derivatives with respect to a quantity add the dimensions of the variable one is differentiating with respect to on the denominator. Thus:

- position (x) has units of L (Length);
- derivative of position with respect to time (dx/dt, velocity) has units of L/T – Length from position, Time from the derivative;
- the second derivative (d^2x/dt^2, acceleration) has units of L/T^2.

In economics, one distinguishes between stocks and flows: a stock has units of "units" (say, widgets or dollars), while a flow is a derivative of a stock, and has units of "units/time" (say, dollars/year).

In some contexts, dimensional quantities are expressed as dimensionless quantities or percentages by omitting some dimensions. For example, Debt to GDP ratios are generally expressed as percentages: total debt outstanding (dimension of Currency) divided by annual GDP (dimension of Currency) – but one may argue that in comparing a stock to a flow, annual GDP should have dimensions of Currency/Time (Dollars/Year, for instance), and thus Debt to GDP should have units of years.

Dimensionless Concepts

Constants: The dimensionless constants that arise in the results obtained, such as the C in the Poiseuille's Law problem and the k in the spring problems discussed above come from a more detailed analysis of the underlying physics, and often arises from integrating some differential equation. Dimensional analysis itself has little to say about these constants, but it is useful to know that they very often have a magnitude of order unity. This observation can allow one to sometimes make "back of the envelope" calculations about the phenomenon of interest, and therefore be able to more efficiently design experiments to measure it, or to judge whether it is important, etc.

Formalisms

Paradoxically, dimensional analysis can be a useful tool even if all the parameters in the underlying theory are dimensionless, e.g., lattice models such as the Ising model can be used to study phase transitions and critical phenomena. Such models can be formulated in a purely dimensionless way. As we approach the critical point closer and closer, the distance over which the variables in the lattice model are correlated (the so-called correlation length, ξ) becomes larger and larger. Now, the correlation length is the relevant length scale related to critical phenomena, so one can, e.g., surmize on "dimensional grounds" that the non-analytical part of the free energy per lattice site should be $\sim 1/\xi^d$ where d is the dimension of the lattice.

It has been argued by some physicists, e.g., Michael Duff, that the laws of physics are inherently dimensionless. The fact that we have assigned incompatible dimensions to Length, Time and Mass is, according to this point of view, just a matter of convention, borne out of the fact that before the advent of modern physics, there was no way to relate mass, length, and time to each other. The three independent dimensionful constants: *c*, ‘, and *G*, in the fundamental equations of physics must then be seen as mere conversion factors to convert Mass, Time and Length into each other. Just as in the case of critical properties of lattice models, one can recover the results of dimensional analysis in the appropriate scaling limit; e.g., dimensional analysis in mechanics can be derived by reinserting the constants ‘, c, and G (but we can now consider them to be dimensionless) and demanding that a nonsingular relation between quantities exists in the limit $c \to \infty$, $\hbar \to 0$ and $G \to 0$. In problems involving a gravitational field the latter limit should be taken such that the field stays finite.

Finance, Economics, and Accounting

In finance, economics, and accounting, dimensional analysis is most commonly referred to in terms of the distinction between stocks and flows. More generally, dimensional analysis is used in interpreting various financial ratios, economics ratios, and accounting ratios.

- For example, the P/E ratio has dimensions of time (units of years), and can be interpreted as "years of earnings to earn the price paid."

- In economics, debt-to-GDP ratio also has units of years (debt has units of currency, GDP has units of currency/year).
- More surprisingly, bond duration also has units of years, which can be shown by dimensional analysis, but takes some financial intuition to understand.
- Velocity of money has units of 1/Years (GDP/Money supply has units of Currency/Year over Currency): how often a unit of currency circulates per year.

Critics of mainstream economics, notably including adherents of Austrian economics, have claimed that it lacks dimensional consistency.

8

Atmospheric Dispersion Modelling

Atmospheric dispersion modelling is the mathematical simulation of how air pollutants disperse in the ambient atmosphere. It is performed with computer programs that solve the mathematical equations and algorithms which simulate the pollutant dispersion. The dispersion models are used to estimate or to predict the downwind concentration of air pollutants or toxins emitted from sources such as industrial plants, vehicular traffic or accidental chemical releases.

Such models are important to governmental agencies tasked with protecting and managing the ambient air quality. The models are typically employed to determine whether existing or proposed new industrial facilities are or will be in compliance with the National Ambient Air Quality Standards (NAAQS) in the United States and other nations. The models also serve to assist in the design of effective control strategies to reduce emissions of harmful air pollutants. During the late 1960s, the Air Pollution Control Office of the U.S. EPA initiated research projects that would lead to the development of models for the use by urban and transportation planners. A major and significant application of a roadway dispersion model that resulted from such research was applied to the Spadina Expressway of Canada in 1971.

Air dispersion models are also used by public safety responders and emergency management personnel for emergency planning of accidental chemical releases. Models are used to determine the consequences of accidental releases of hazardous or toxic materials, Accidental releases may result fires, spills or explosions that involve hazardous materials, such as chemicals or radionuclides. The results of dispersion modelling, using worst case accidental release source terms and meteorological conditions, can provide an estimate of location impacted areas, ambient concentrations, and be used to determine protective actions appropriate in the event a release occurs. Appropriate protective actions may include evacuation or shelter in place for persons in the downwind direction. At industrial facilities, this type of consequence assessment or emergency planning is required under the Clean Air Act (United States) (CAA) codified in Part 68 of Title 40 of the Code of Federal Regulations.

The dispersion models vary depending on the mathematics used to develop the model, but all require the input of data that may include:

- Meteorological conditions such as wind speed and direction, the amount of atmospheric turbulence (as characterized by what is called the "stability class"), the ambient air temperature, the height to the bottom of any inversion aloft that may be present, cloud cover and solar radiation.
- Source term (the concentration or quantity of toxins in emission or accidental release source terms) and temperature of the material
- Emissions or release parameters such as source location and height, type of source (i.e., fire, pool or vent stack) and exit velocity, exit temperature and mass flow rate or release rate.
- Terrain elevations at the source location and at the receptor location(s), such as nearby homes, schools, businesses and hospitals.
- The location, height and width of any obstructions (such as buildings or other structures) in the path of the emitted gaseous plume, surface roughness or the use of a more generic parameter "rural" or "city" terrain.

Many of the modern, advanced dispersion modelling programs include a pre-processor module for the input of meteorological and other data, and many also include a post-processor module for graphing the output data and/or plotting the area impacted by the air pollutants on maps. The plots of areas impacted may also include isopleths showing areas of minimal to high concentrations that define areas of the highest health risk. The isopleths plots are useful in determining protective actions for the public and responders.

The atmospheric dispersion models are also known as atmospheric diffusion models, air dispersion models, air quality models, and air pollution dispersion models.

Atmospheric Layers

Discussion of the layers in the Earth's atmosphere is needed to understand where airborne pollutants disperse in the atmosphere. The layer closest to the Earth's surface is known as the *troposphere.* It extends from sea-level to a height of about 18 km and contains about 80 percent of the mass of the overall atmosphere. The *stratosphere* is the next layer and extends from 18 km to about 50 km. The third layer is the *mesosphere* which extends from 50 km to about 80 km. There are other layers above 80 km, but they are insignificant with respect to atmospheric dispersion modelling.

The lowest part of the troposphere is called the *atmospheric boundary layer (ABL)* or the *planetary boundary layer (PBL)* and extends from the Earth's surface to about 1.5 to 2.0 km in height. The air temperature of the atmospheric boundary layer decreases with increasing altitude until it reaches what is called the *inversion layer* (where the temperature increases with increasing altitude) that caps the atmospheric boundary layer. The upper part of the troposphere (i.e., above the inversion layer) is called the *free troposphere* and it extends up to the 18 km height of the troposphere.

The ABL is of the most important with respect to the emission, transport and dispersion of airborne pollutants. The part of the ABL between the Earth's surface and the bottom of the inversion layer is known as the mixing layer. Almost all of the airborne pollutants emitted into the ambient atmosphere are transported and dispersed within the mixing layer. Some of the emissions penetrate the inversion layer and enter the free troposphere above the ABL.

In summary, the layers of the Earth's atmosphere from the surface of the ground upwards are: the ABL made up of the mixing layer capped by the inversion layer; the free troposphere; the stratosphere; the mesosphere and others. Many atmospheric dispersion models are referred to as *boundary layer models* because they mainly model air pollutant dispersion within the ABL. To avoid confusion, models referred to as *mesoscale models* have dispersion modelling capabilities that extend horizontally up to a few hundred kilometres. It does not mean that they model dispersion in the mesosphere.

Breakwater (Structure)

Breakwaters are structures constructed on coasts as part of coastal defence or to protect an anchorage from the effects of both weather and longshore drift.

Purposes of Breakwaters

Offshore breakwaters, also called bulkhead, reduce the intensity of wave action in inshore waters and thereby reduce coastal erosion or provide safe harborage. Breakwaters may also be small structures designed to protect a gently sloping beach and placed one to three hundred feet offshore in relatively shallow water.

An anchorage is only safe if ships anchored there are protected from the force of high winds and powerful waves by some large underwater barrier which they can shelter behind. Natural harbours are formed by such barriers as headlands or reefs. Artificial harbors can be created with the help of breakwaters. Mobile harbours, such as the D-Day Mulberry harbours, were floated into position and acted as breakwaters. Some natural harbours, such as those in Plymouth Sound, Portland Harbour and Cherbourg, have been enhanced or extended by breakwaters made of rock.

Unintended Consequences

The dissipation of energy and relative calm water created in the lee of the breakwaters often encourage accretion of sediment (as per the design of the breakwater scheme). However this can lead to excessive salient build up, leading to tombolo formation reducing longshore drift shoreward of the breakwaters (Sea Palling, UK). This trapping of sediment can cause adverse effects down drift of the breakwaters leading to beach sediment starvation and increased erosion. This may then lead to further engineering protection being needed down drift of the breakwater development.

Breakwaters are subject to damage, and overtopping in severe storms events.

Types of Breakwater Structures

A breakwater structure is designed to absorb the energy of the waves that hit it, either by using mass (e.g., with caissons), or by using a revetment slope (e.g., with rock or concrete armour units).

In Coastal Engineering, a revetment is a land backed structure whilst a breakwater is a sea backed structure (i.e., water on both sides).

Caisson breakwaters typically have vertical sides and are usually erected where it is desirable to berth one or more vessels on the inner face of the breakwater. They use the mass of the caisson and the fill within it to resist the overturning forces applied by waves hitting them. They are relatively expensive to construct in shallow water, but in deeper sites they can offer a significant saving over revetment breakwaters.

Rubble mound breakwaters use structural voids to dissipate the wave energy. Rock or concrete armour units on the outside of the structure absorb most of the energy, while gravels or sands prevent the wave energy's continuing through the breakwater core. The slopes of the revetment are typically between 1:1 and 1:2, depending upon the materials used. In shallow water, revetment breakwaters are usually relatively inexpensive. As water depth increases, the material requirements, and hence costs, increase significantly.

Clapotis

In hydrodynamics, the clapotis (from French: *"lapping of water"*) is a non-breaking standing wave pattern, caused for example, by the reflection of a travelling surface wave train from a near vertical shoreline like a breakwater, seawall or steep cliff. The resulting clapotic wave does not travel horizontally, but has a fixed pattern of nodes and antinodes. These waves promote erosion at the toe of the wall, and can cause severe damage to shore structures. The term was coined in 1877 by French mathematician and physicist Joseph Valentin Boussinesq who called these waves 'le clapotis' meaning 'standing waves'.

In the idealized case of "full clapotis" where a purely monotonic incoming wave is completely reflected normal to a solid vertical wall, the standing wave height is twice the height of the incoming waves at a distance of one half wavelength from the wall. In this case, the circular orbits of the water particles in the deep-water wave are converted to purely linear motion, with vertical velocities at the antinodes, and horizontal velocities at the nodes. The standing waves alternately rise and fall in a mirror image pattern, as kinetic energy is converted to potential energy, and vice versa. In his 1907 text, *Naval Architecture*, Cecil Peabody described this phenomenon:

At any instant the profile of the water surface is like that of a trochoidal wave, but the profile instead of appearing to run to the right or left, will grow from a horizontal surface, attain a maximum development, and then flatten out till the surface is again horizontal; immediately another wave profile will form with its crests where the hollows formerly were, will grow and flatten out, etc. If attention is concentrated on a certain crest, it will be seen to grow to its greatest height, die away, and be succeeded in the same place by a hollow, and the interval of time between the successive formations of crests at a given place will be the same as the time of one of the component waves."

Related Phenomena

True clapotis is very rare, because the depth of the water or the precipitousness of the shore are unlikely to completely satisfy the idealized requirements. In the more realistic case of partial clapotis, where some of the incoming wave energy is dissipated at the shore, the incident wave is less than 100% reflected, and only a partial standing wave is formed where the water particle motions are elliptical. This may also occur at sea between two different wave trains of near equal wavelength moving in opposite directions, but with unequal amplitudes. In partial clapotis the wave envelope contains some vertical motion at the nodes. When a wave train strikes a wall at an oblique angle, the reflected wave train departs at the supplementary angle causing a cross-hatched wave interference pattern known as the clapotis gaufré ("waffled clapotis"). In this situation, the individual crests formed at the intersection of the incident and reflected wave train crests move parallel to the structure. This wave motion, when combined with the resultant vortices, can erode material from the seabed and transport it along the wall, undermining the structure until it fails.

Clapotic waves on the sea surface also radiate infrasonic microbaroms into the atmosphere, and seismic signals called microseisms coupled through the ocean floor to the solid Earth.

Iribarren Number

In fluid dynamics, the Iribarren number or Iribarren parameter – also known as the surf similarity parameter and breaker parameter – is a dimensionless parameter used to model several effects of (breaking) surface gravity waves on beaches and coastal structures. The parameter is named after the Spanish engineer Ramón Iribarren Cavanillas (1900–1967), who introduced it to describe the occurrence of wave breaking on sloping beaches.

For instance, the Iribarren number is used to describe breaking wave types on beaches; or wave run-up on – and reflection by – beaches, breakwaters and dikes.

Definition

The Iribarren number – often denoted as *Ir* or ξ – is defined as:

$$\xi = \frac{\tan\alpha}{\sqrt{H / L_0}}, \quad \text{with} \quad L_0 = \frac{g}{2\pi}T^2,$$

where ξ is the Iribarren number, α is the bed slope, H is the wave height, L_0 is the deep-water wavelength, T is the period and g is the gravitational acceleration. Depending on the application, different definitions of H and T are used, for example: for periodic waves the wave height H_0 at deep water or the breaking wave height H_b at the edge of the surf zone. Or, for random waves, the significant wave height H_s at a certain location.

Hudson's Equation

Hudson's Equation, also known as Hudson's Formula, is an equation used by coastal engineers to calculate the minimum size of riprap (rock armour blocks) required to provide *satisfactory* stability characteristics for rubble structures such as breakwaters under attack from storm wave conditions. The equation was developed by the United States Army Corps of Engineers, Waterways Experiment Station (WES), following extensive investigations by Hudson (1953, 1959, 1961a, 1961b).

Initial Equation

The equation itself is:

$$W = \frac{\gamma_r H^3}{K_D \Delta^3 \cot\theta}$$

where:

- W is the design weight of the riprap armour (Newton)
- γ_r is the specific weight of the armour blocks (N/m^3)
- H is the design wave height at the toe of the structure (m)
- K_D is a dimensionless stability coefficient, deduced from laboratory experiments for different kinds of armour blocks and for very small damage (a few blocks removed from the armour layer) (-):
- K_D = around 3 for natural quarry rock
- K_D = around 10 for artificial interlocking concrete blocks
- Δ is the dimensionless relative buoyant density of rock, i.e. $(\rho_r / \rho_w - 1)$ = around 1.58 for granite in sea water
- ρ_r and ρ_w are the densities of rock and (sea)water (-)
- θ is the angle of revetment with the horizontal

Updated Equation

This equation was rewritten as follows in the nineties:

$$\frac{H_s}{\Delta D_{n50}} = (K_D cot\theta)^{1/3}$$

where

- H_s is the design significant wave height at the toe of the structure (m)
- Δ is the dimensionless relative buoyant density of rock, i.e. $(\rho_r / \rho_w - 1)$ = around 1.58 for granite in sea water
- ρ_r and ρ_w are the densities of rock and (sea)water (-)
- D_{n50} is the nominal median diameter of armour blocks = $(W_{50}/\rho_r)^{1/3}$ (m)
- K_D is a dimensionless stability coefficient, deduced from laboratory experiments for different kinds of armour blocks and for very small damage (a few blocks removed from the armour layer) (-):
- K_D = around 3 for natural quarry rock
- K_D = around 10 for artificial interlocking concrete blocks
- θ is the angle of revetment with the horizontal

The armour blocks may be considered stable if the *stability number* $N_s = H_s / \Delta D_{n50} < 1.5$ to 2, with damage rapidly increasing for $N_s > 3$.

Obviously, these equations may be used for preliminary design, but scale model testing (2D in wave flume, and 3D in wave basin) is absolutely needed before construction is undertaken.

Revetment

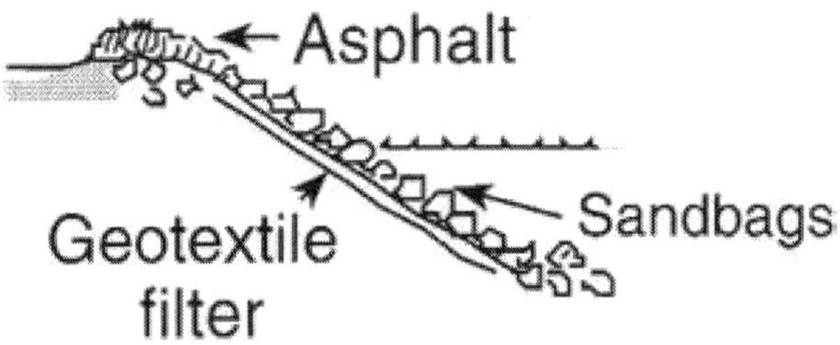

Figure: *Asphalt and sandbag revetment with a geotextile filter*

In stream restoration, river engineering or coastal management, revetments are sloping structures placed on banks or cliffs in such a way as to absorb the energy of incoming water. In military engineering they are structures, again sloped, formed to secure an area from artillery, bombing, or stored explosives. River or coastal revetments are usually built to preserve the existing uses of the shoreline and to protect the slope, as defence against erosion.

Freshwater Revetments

Any revetments are used to line the banks of freshwater rivers, lakes, and man-made reservoirs, especially to prevent damage during periods of floods or heavy seasonal rains. Many materials may be used: wooden piles, loose-piled boulders or concrete shapes, or more solid banks.

Revetments as Coastal Defence

Revetments are used as a low-cost solution for coastal erosion defence in areas where crashing waves may otherwise deplete the coastline. Wooden revetments are made of planks laid against wooden frames so that they disrupt the force of the water. Although once popular, the use of wooden revetments has largely been replaced by modern concrete-based defence structures such as tetrapods.

Tetrapods

In coastal engineering, a tetrapod is a four-legged concrete structure used as armour unit on breakwaters. The tetrapod's shape is designed to dissipate the force of incoming waves by allowing water to flow around rather than against it, and to reduce displacement by allowing a random distribution of tetrapods to mutually interlock. Tetrapods have been shown to offer little advantage compared to other concrete armour units. Layering and the amount of tetrapods have shown to have no appreciable effect on the stability of the tetrapod structures although stability increases with the size of the tetrapod pad. Tetrapods have also been criticized for causing more damage than they prevent because they alter ocean currents and disrupt the natural cycles of erosion and deposition that form and reshape coasts. Concrete coastal installations can also be lethally dangerous to swimmers and surfers, as well to shipping and recreational boaters.

Fortifications

According to the U.S. National Park Service, and referring mostly to their employment in the American Civil War, a revetment is defined as a "retaining wall constructed to support the interior slope of a parapet. Made of logs, wood planks, fence rails, fascines, gabions, hurdles, sods, or stones, the revetment provided additional protection from enemy fire, and, most importantly, kept the interior slope nearly vertical. Stone revetments commonly survive. A few log revetments have been preserved due to high resin pine or cypress and porous sandy soils. After an entrenchment was abandoned, many log or rail revetments were scavenged for other uses, causing the interior slope to slump more quickly. An interior slope will appear more vertical if the parapet eroded with the revetment still in place."

Sand Dune Stabilization

Sand dunes are common features of shoreline and desert environments. Dunes provide habitat for highly specialised plants and animals, including rare and endangered species. They can protect beaches from erosion and recruit sand to eroded beaches. Dunes are threatened by human activity, both intentional and unintentional. Countries such as the United States, Australia, Canada, New Zealand, the United Kingdom, and Netherlands, operate significant dune protection programs. Stabilizing dunes involves multiple actions. Planting vegetation reduces the impact of wind and water. Fences catch sand and other material. Footpaths protect dunes from damage from foot traffic.

Vegetation

The location of the dune limits the types of plant that can thrive there. Beach dunes consist of th *foredune*, the angled side which faces the ocean, the *sand plain* at the top of the dune, which may or may not be pesent, and the *backdune*, the angled side that faces away from the ocean.

Foredune Flora

Plants that thrive on the foredune must be tolerant to salt spray, strong winds, and burial by blowing/accumulating sand. Typical vegetation includes Ammophila arenaria, Honckenya peploides, Cakile maritima, and Spartina coarctata.

Backdune Flora

Plants which thrive on the broad dune plain and backdune grow together into dense patches termed *dune mats* that hold the dune together. Vegetation typical of the plain and backdune include Hudsonia tomentosa, Spartina patens, Iva imbricata, and Eregeron glaucus. Introduced species can outcompete native plants and disrupt animal life, making them formally "invasive species".

Shrub Stage

The above species are herbaceous plants. After they have rooted and developed fully, a second stage, the "shrub stage", can begin. During this phase, larger plants with deeper root systems can be planted. Examples are Empetrum nigrum, Ilex vomitoria, and Vaccinium ovatum. The shrub stage is usually the final phase and may last for short or long periods of time depending on microclimatic conditions such as distance from the shoreline, availability of groundwater, or salt spray effects..

Coastal Management

A single beach may be divided into segments with different owners. Gaining agreement among all owners complicates the process of stabilizing the dunes. Without agreement, some parts of the dune may go unplanted, while others host visually appealing plants that do little to stabilize the dune. Inconsistent vegetation, known as a fragmened or decoupled gradient, can create weak points in the dune that limit its effectiveness against floods and even its continued existence.

Publicly owned beaches, found in U.S. states such as California and Hawaii and in other jurisdictions, present the opportunity to systematically manage—or mismanage—beaches and their accompanying dune.

Human Impacts

It is generally observed that coastal dunes are highly sensitive to any number of human activities. It is also understood that coastal dunes need to have limited interaction with humans in order for their protection and survival. Around the world coastal dunes are eroding and under pressure due to increased sediment loss because of increased coastal human activities. For example, in the last 30 years, mainly because of tourism, nearly 75% of the Mediterranean's coastal dunes have been damaged or destroyed. Human impactscan expand from minor to highly significant adverse effects. Minor effects include pedestrian trampling, constructing footpaths, and off road vehicle tracks, to the highly more significant which includes the construction of roads, car parks, houses, and the use of sand dunes for ground water extraction, the conversion of sand dunes to land for farming, forestry, waste disposal and sand dune mining.

New Zealand is a country with a very long coast line that consists of around 1090 km of sand dunes. From the early 1900s to the early 2000s the area of sand dunes has declined by about 70% from 129,000 ha to 39,000 ha. This decline has been mainly due to erosion caused by the depletion of vegetation cover in order to convert sand dunes to land for forestry, farming and urban areas. Loss in vegetation cover exposes sand to the action of wind and leaves sand dunes vulnerable to increased erosion as it is vegetation that binds sand together. The loss of vegetation causes sand to move inland which is a process known as sand encroachment, this process can having damaging effects on inland properties and leaves properties vulnerable to future coastal hazards for example storm surge.

As well as human activities, natural processes operate within coastal dune environments, these processes include events such as storms, floods, sea level change, and sediment supply changes. Coastal dune systems are mainly controlled by a principle natural process called the sediment supply and transport system. This system consists of an offshore zone (sediment bank), transit zone (beach and active foredune) and a resting zone (stable sand dunes). The sediment in this system is supplied by long shore drift predominantly from rivers and from the erosion of cliffs, headlands, other dune systems and when sediment supply is low sand dunes are subject to erosion. Humans affect sediment supply in a number of ways including the damming of rivers, dredging in coastal water and the construction of harbours, all of which depletes sediment supply. They also disrupt the long shore transport of sediment through the construction of piers, marinas or groynes, breakwaters and other hard structures. Excavating, building roads, houses and walkways on coastal dunes or beaches also causes sand dune erosion by disrupting natural sedimentprocesses. Effective management of coastal dune systems must consider natural processes as a normal part of the natural environment as well as controlling human activities and seeking to alter natural processes should be kept at a minimum in order for restoration to be successful.

Integrated Management

Sand dunes are an important part of the coast as they are land's first defence against the forces of the ocean. Because of this, management of sand dunes will focus on the natural authenticity and values of the dunes themselves including flora, fauna, and the ecosystem as a whole. The concept of integrated management can be incorporated into management of coastal sand dunes. This concept is made up of three main aspects:

- Restore and conserve natural processes
- Create conditions for future natural processes
- Develop types of land use that agree with the above

These concepts seek to incorporate natural processes and human activities within coastal dunes systems. Sand dune management issues are related to the nature of sediment, which has no cohesiveness so the influence of wind and sediment supply is an important factor in determining erosion and deposition of dry sand grains .

Sand dune management consists of restoring natural dune systems and processes with the implementation of "soft" technique which includes the cessation of any activity that adversely effects the sediment supply of the dunes, planting vegetation, fencing off sensitive areas, creating dune walkways, and providing information to the public regarding the importance of maintaining stable sand dunes. The concept of "soft" techniques is in contrast to the widely used "hard" techniques which consist of protecting the coast through the construction of concrete sea walls, revetments, gabions, groynes and offshore break waters. The problem with using these "hard" engineering methods is that they only provide temporary protection against ocean hazards and each "hard" method produces an "end effect" that causes erosion problems to adjacent coastal areas. This is why the implementation of "soft" techniques in dune restoration should be used more widely in coastal protection as they provide natural, long term protection with no adverse effect to adjacent coastal areas. The use of soft techniques can be seen in many coastal protection projects around the world including many in New Zealand.

Restoration and Protection

One of the first things a coastal manager must look for when establishing sand dune management is to ascertain the main cause of sediment loss and from there, make decisions on how to minimise the cause and how to restore sand dunes appropriately. In New Zealand the main cause of dune erosion has been the loss of sand vegetation and this means that sand dune management consists primarily of stabilizing dune systems through the "soft" technique of re planting and protecting native dune vegetation. It is advised that dunes should not be restored with the planting of exotic species and managers should aim to only use species indigenous to an area, originally New Zealand used exotic, fast growing dune vegetation called Marram Grass) to stabilize eroding sand dunes. It was later discovered that this exotic species created very high, steep sand dunes that were increasing susceptible to New Zealand's high storm energy climate, it also became apparent that the native New Zealand dune vegetation were out competed by this introduced species which caused issues biologically. At present there is a large push to eradicate Marram grass as it is now classed as a pest species in New Zealand and at present coastal managers encourage the planting of the two indigenous species, Pingao (Desmoschoenus spiralis) and Spinifex (Spinifex sericeus) as these two species create low angle, strong, stable dunes.

The management of dune restoration projects includes planting vegetation successfully. Merely scattering and leaving seeds over sandy areas is perceived as a potential waste as strong winds can blow seeds away. Planting of nurseries should be conducted in the hollows of any present sand hills or in depressions of sandy areas. Seeds should be planted in moist conditions and should be lightly covered by a layer of sand and large mature grass individuals should be planted in deep holes with the tops of the plants cut short so as to reduce wind resistance.

Planting of sand dune vegetation alone does not entirely restore or protect sand dunes from human impacts. Seedlings need to be protected from human trampling which is done mainly by fencing off newly planted areas. Fencing also acts as another way of trapping and accumulating sand among the dunes and can also act as protection against strong winds. Figure two shows an area of newly planted vegetation at Spencer Park Beach, New Zealand, it shows that fencing protects plants from trampling and figure three shows how small fencing using chicken wire can encourage plants to grow. In addition to erecting fences to protect sand dune vegetation, the establishment of walkways will protect developed sand dunes from degrading again overtime. These walkways provide access to beaches for the public that does not require pedestrians to cross the dunes themselves thus preventing damage. Figure four shows a typical walkway at New Brighton Beach, New Zealand, the walk way itself is built to go over the sand dunes as opposed to straight through them, in order to maintain the continuity and strength of the dunes.

The Public

Because sand dunes and beaches are extremely popular for outdoor recreation, the success of restoration and dune management depends a great deal on the information available and understanding of the public. This is important because not all aspects of sand dune management are obvious or important to the public if they do not understand why management is necessary e.g. stabilization. There are many ways in which information can be provided to the public including the erection of panels at entrances to dune walkways, leaflets, brochures, maps and council workers who can talk to the public about the importance of protecting sand dunes. Figure five shows a photo of a panel put up at Spencer Park by the Christchurch City Council indicating that people should not walk on the area, as it is an area for sand dune restoration. It needs to be re-iterated that sand dune restoration and protection is in the public's best interest as the loss of sediment on any coast line leads to the erosion of beaches, which in turn leads to a loss of recreational value of many coast lines. The public needs to be aware of the role that sand dunes play in coastal protection because without public support coastal sand dune management will be virtually impossible.

Sand Engine

The sand engine is an experiment in the management of dynamic coastline. It is run off South Holland. A sandbar-shaped peninsula was created by man; the surface is about 1 km^2. It is expected that this sand is then moved over the years by the action of waves, wind and currents along the coast. For the first time in 2011 at the request of the Hoogheemraadschap van Delfland as part of the coastal management and the maintenance of the coastline by the highways and the province of South Holland, a peninsula was created between Ter Heijde and Kijkduin, where natural beaches and dunes are relatively narrow.

Building with Nature

Marcel Stive may be considered as the father of this project. Another such project could provide a solution to the weak coast between Camperduin and Petten, which is called the Hondsbossche seawall. The experience of this first sand enine is interesting for future projects. Th construction of the sand engine at Ter Heijde cost 70 million euros and is the first in the world of its kind. Joop Atsma, State Secretary for Infrastructure and Environment, presented the project in November 2011 and his purpose was to convince that the technique of sand engine could be useful on more locations along the Dutch coast.

Delta Dune

The first such project was given the name *Delta Dune.* Work began in January 2011 the conditions were favourable and the operation was completed in October. A volume of 21 million cubic metres of sand covers an area of 128 ha.

Trade-Offs

A cost benefit approach is an effective way to determine whether a seawall is appropriate and whether the benefits are worth the expense. Besides controlling erosion, consideration must be given to the effects of hardening a shoreline on natural coastal ecosystems and human property or activities. A seawall is a static feature which can conflict with the dynamic nature of the coast and impede the exchange of sediment between land and sea. The table below summarises some positive and negative effects of seawalls which can be used when comparing their effectiveness with other coastal management options, such as beach nourishment.

Advantages and disadvantages of seawalls according to Short (1999)	
Advantages	***Disadvantages***
• Long term solution in comparison to soft beach nourishment. • Effectively minimizes loss of life in extreme events and damage to property caused by erosion. • Can exist longer in high energy environments in comparison to 'soft' engineering methods. • Can be used for recreation and sightseeing. • Forms a hard and strong coastal defence.	• Very expensive to construct. • Can cause beaches to dissipate rendering them useless for beach goers. • Scars the very landscape that they are trying to save and provides an 'eyesore.' • Reflected energy of waves leading to scour at base. • Can disrupt natural shoreline processes and destroy shoreline habitats such as wetlands and intertidal beaches. • Altered sediment transport processes can disrupt sand movement that can lead to increased erosion down drift from the structure.

Generally seawalls can be a successful way to control coastal erosion, but only if they are constructed well and out of materials which can withstand the force of ongoing wave energy. Some understanding is needed of the coastal processes and morphodynamics specific to the seawall location. Seawalls can be very helpful; they can offer a more long term solution than soft engineering options, additionally providing recreation opportunities and protection from extreme events as well as everyday erosion. Extreme natural events expose weaknesses in the performance of seawalls, and analyses of these can lead to future improvements and reassessment.

Simulations

In 2007 researchers at the University of Salerno published studies showing interactions between maritime breakwaters and waves using CAD and CFD software. In the simulations the filtration motion of the fluid within the interstices, which normally exist in a breakwater, is estimated by integrating the relevant RANS equations coupled with a random number generated turbulence model inside the voids, rather than using the classical equations for porous media. The breakwaters were modelled, both for the actual size construction and for a physical laboratory test, by overapping three-dimensional elements. The numerical grid was thickened in such a way to have some computational nodes along the flow paths among the breakwater's blocks.

Issues

Sea Level Rise: Sea level rise creates an issue for seawalls worldwide as it raises both the mean normal water level and the height of waves during extreme weather events, which the current seawall heights may be unable to cope with (Allan *et al.* 1999). The International Panel on Climate Change (IPCC) (1997) suggested that sea level rise over the next 50 – 100 years will accelerate with a projected increase in global mean sea level of +18 cm by 2050 AD. This data is reinforced by Hannah (1990) who calculated similar statistics including a rise of between +16-19.3 cm throughout 1900–1988. This problem could be overcome by further modelling and determining the extension of height and reinforcement of current seawalls which needs to occur for safety to be ensured in both situations.

Extreme Events

Extreme events also pose a problem as it is not easy for people to predict or imagine the strength of hurricane or storm induced waves compared to normal, expected wave patterns. An extreme event can dissipate hundreds of times more energy than everyday waves, and calculating structures which will stand the force of coastal storms is difficult and, often the outcome can become unaffordable. For example, Omaha Beach seawall in New Zealand was designed to prevent erosion from everyday waves only, and when a storm in 1976 carved out 10m behind the existing seawall the whole structure was destroyed (GeoResources, 2001).

Other Issues

Some Further Issues Include: lack of long term trend data of seawall effects due to a relatively short duration of data records; modelling limitations and comparisons of different projects and their effects being invalid or unequal due to different beach types; materials; currents; and environments (Christchurch City Council, 2009).

Historical Examples

Seawall construction has existed since ancient times. In the 1st century BCE, Romans built a seawall / breakwater at Caesarea Maritima creating an artificial harbor (Sebastos Harbor). The construction used Pozzolana concrete which hardens in contact with sea water. Barges were constructed and filled with the concrete. They were floated into position and sunk. The resulting harbor / breakwater / sea wall is still in existence today - more than 2000 years later.

More recently, sea walls were constructed in 1623 in Canvey Island, UK, when great floods of the Thames estuary occurred, prompting the construction of protection for further events in this flood prone area (Council of Europe, 1999). Since then, seawall design has become more complex and intricate in response to an improvement in materials, technology and an understanding of how coastal processes operate. This section will outline some key case studies of seawalls in chronological order and describe how they have performed in response to tsunami or ongoing natural processes and how effective they were in these situations. Analysing the successes and shortcomings of seawalls during severe natural events allows their weaknesses to be exposed, and areas become visible for future improvement.

Pondicherry

On December 26, 2004, towering waves of the 2004 Indian Ocean earthquake tsunami crashed against India's south-eastern coastline killing thousands. However, the former French colonial enclave of Pondicherry (now Pondicherry) escaped unscathed. This was primarily due to French engineers who had constructed (and maintained) a massive stone seawall during the time which the city was a French colony. This 300 year old seawall effectively kept Pondicherry's historic centre dry even though tsunami waves drove water 24 feet above the normal high-tide mark.

The barrier was initially completed in 1735 and over the years, the French continued to fortify the wall, piling huge boulders along its 1.25 mile (2 km) coastline to stop erosion from the waves pounding the harbour. At its highest, the barrier running along the water's edge reaches about 27 feet above sea level.

The boulders, some weighing up to a ton, are weathered black and brown. The sea wall is inspected every year and whenever gaps appear or the stones sink into the sand, the government adds more boulders to keep it strong (Allsop, 2002).

The Union Territory of Pondicherry recorded some 600 deaths from the huge tsunami waves that struck India's coast after the mammoth underwater earthquake (which measured 9.0 on the moment magnitude scale) off Indonesia, but most of those killed were fishermen who lived in villages beyond the artificial barrier which reinforces the effectiveness of seawalls.

Vancouver

The Vancouver Seawall is a stone seawall constructed around the perimeter of Stanley Park in Vancouver. The seawall was constructed initially as waves created by ships passing through the First Narrows were eroding the area between Prospect Point and Brockton Point. The Vancouver Seawall also exemplfies how seawalls can be utilised and valued for recreational activities and coastal sightseeing. A pedestrian, cycling and roller blading pathway exists on the seawall and has been extended far outside the parameters of Stanley Park.

Construction of the seawall began in 1917, and since then this pathway has become one of the most used features of the park by both locals and tourists and now extends 22 km in total (Belyea & Ross, 1992). The construction of the seawall also provided employment for relief workers during the Great Depression and seamen from HMCS *Discovery* on Deadman's Island who were facing punishment detail in the 1950s (Steele, 1985).

Overall, the Vancouver Seawall is a prime example of how seawalls can simultaneously provide shoreline protection and a source of recreation which enhances human enjoyment of the coastal environment. It also illustrates that although shoreline erosion is a natural process, human activities, interactions with the coast and poorly planned shoreline development projects can accelerate natural erosion rates.

9

Multiple Force Members

The term, *three-force member,* is often incorrectly applied to members with more than three forces. Any member which is subjected to more than three single concentrated loads should be identified as a Multiple Force Member. The "extra" forces acting upon a multiple force member can be replaced until the total number of forces is reduced to three; the resultant force of the loading and the two reactions. This means that distributed loads must always be replaced so that only one resultant force is applied to the member. Then, all of the other forces must be replaced so that there is only one resultant representing the external loading. The three non-parallel forces must be concurrent so that the three-force principle then applies.

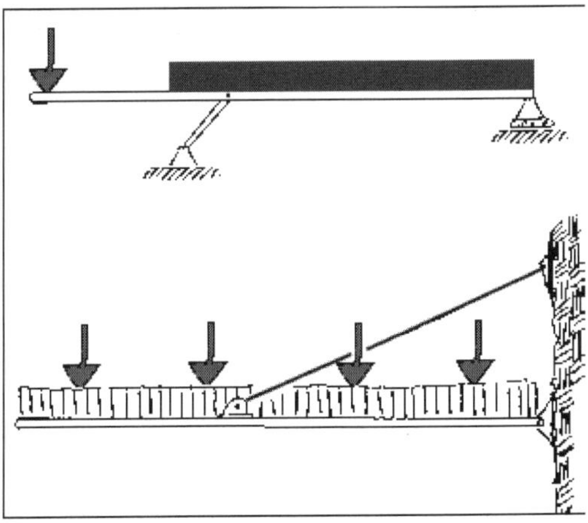

The loading of the first image would be resolved in the following manner. The resultant of the distributed load would be determined. This would act at the midpoint of the width of the load. The resultant of this force and the concentrated load acting at the cantilevered end of the beam would be combined into one load by summing moments about the right-hand support. This system would then be reduced from multiple force member to a three force member.

The second system could represent an entrance canopy to a hotel that also has four large spot lights sitting upon it. These six loads must also be combined into one load resultant before the member can be considered a three force member.

Primary Loads

The loading of a building structure can take on a wide variety of forms. In many cases the exact loading will not fit neatly into a specific category. Yet, loads can usually be considered to be Primary or Secondary. Primary loads are discussed below. Secondary loads are those loads due to temperature changes, construction eccentricities, shrinkage of structural materials, settlement of foundations, or other such loads.

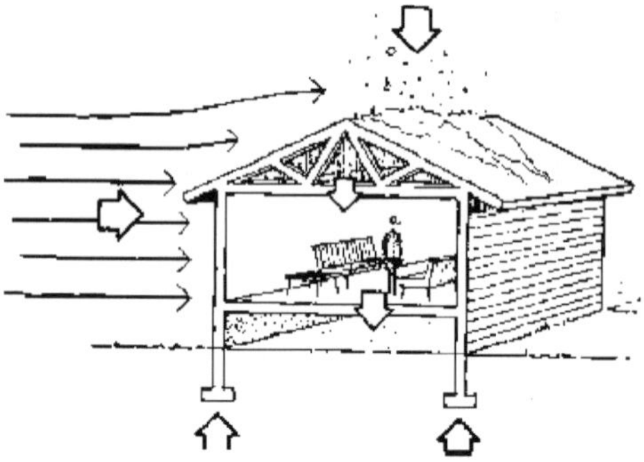

Despite the fact that each and every load and loading combination should be considered in order to reduce the chance of structural failure, the determination of the loading remains a statistical exercise. Each and every load cannot be foreseen; thus, it is critical to determine the worst case that is reasonable to assume to act upon the structure. The sources of primary loading include the materials from which the structure was built, the occupants, their furniture, and various weather conditions, as well as unique loading conditions experienced during construction, extreme weather and natural catastrophes. Primary Loads are divided into two broad categories according to the way in which they act upon the structure or structural element. These are DEAD LOADS and LIVE LOADS. When considering the possible combinations of these two categories of loading, the odds of certain loads occurring simultaneously are assumed to be null. One such combination would be heavy snow, a typhoon, a raging fire and an earthquake. It is possible that two of the first three could occur as the earth quakes, but not that all four would be present at the same time. Thus, one need only consider reasonable loading combinations.

Dead Loads

Dead Loads are those loads which are considered to act permanently; they are "dead," stationary, and unable to be removed. The self-weight of the structural members normally provides the largest portion of the dead load of a building. This will clearly vary with the actual materials chosen. Permanent non-structural elements such as roofing, concrete, flooring, pipes, ducts, interior partition walls, Environmental Control Systems machinery, elevator machinery and all other construction systems within a building must also be included in the calculation of the total dead load. These loads are represented by the red arrow in the illustration.

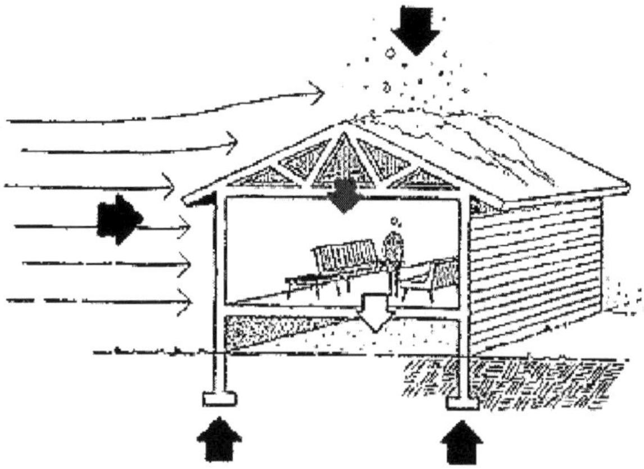

The magnitude of the dead loads of a building can usually be determined with only a 5% margin of error. Properties of building materials are often tabulated and published by authors of textbooks and by the manufacturers. It is always very important to be sure to keep up-to-date on the changes in building materials. Properties of construction materials often vary due to the rapidly changing marketplace. The load due to these materials is often expressed as kN/m^3 or lbs/ft^3. These are normally converted to load/area or load/ length for further consideration.

Live Loads

Live Loads are not permanent and can change in magnitude. They include items found within a building such as furniture, pianos, safes, people, books, cars, computers, machinery, or stored materials, as well as environmental effects such as loads due to the sun, earth or weather. Wind and earthquakes loads are put into the special category of lateral live loads due to the severity of their action upon a building and their potential to cause failure.

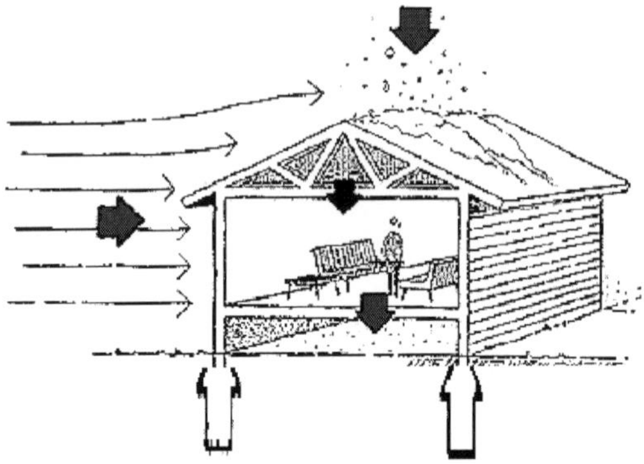

Most buildings have a working life that extends beyond the initially perceived use of the building. Thus, it is almost impossible to predict all of the potential uses that any given structure will experience before it is demolished. If, and when, a building is to be used for a purpose other than its original design, the capacity of the building for its new use must be determined. Since the body of knowledge about the behaviour of buildings is always increasing, a building which might have been designed according to the latest loading information in one year, might not satisfy the requirements a few years later. This has especially been the case with the effects of earthquake loading.

The magnitudes of live loads are difficult to determine with the same degree of accuracy that is possible with dead loads. The probable maximum value of live loads has been determined by research and is included in national building codes. These are usually a minimum design load per area. Building codes also provide for load reductions under certain conditions. As an example, full live loads will not occur on every floor of a multi-story building at the same time. Therefore, the design live load for some of the columns and the foundation can be reduced. Building codes around the world do not concur on the magnitude of the appropriate design live load values. It is critical that the designer take the time to determine the values set down in the local building codes. These are legal documents and MUST be followed.

Lateral Loads

Most lateral loads are live loads whose main component is a horizontal force acting on the structure. Typical lateral loads would be a wind load against a facade, an earthquake, the earth pressure against a beach front retaining wall or the earth pressure against a basement wall. Most lateral loads vary in intensity depending on the building's geographic location, structural materials, height and shape. The dynamic effects of wind and earthquake loads are usually analysed as an equivalent static load in most small and moderate-sized buildings. Others must utilise the iterative potential of the computer. The design wind and earthquake loads on a building are substantially more complex than the following brief discussion and simple examples would indicate. The Uniform Building Code describes the design wind load determination in more detail for the various parts of the United States.

Wind Loads

The most common lateral load is a wind load. The Eiffel Tower is one example of a building which has a structure that was designed to resist a high wind load. Wind against a building builds up a positive pressure on the windward side and a negative pressure (or suction) on the leeward side. Depending upon the shape of the structure it may also cause a negative pressure on the side walls or even the roof. The pressure on the walls and roof is not uniform, but varies across the surface. Winds can apply loads to structures from unexpected directions. Thus, a designer must be well aware of the dangers implied by this lateral load. The magnitude of the pressure that acts upon the surfaces is proportional to the square of the wind speed.

Wind loads vary around the world. Meteorological data collected by national weather services are one of the most reliable sources of wind data. Factors that effect the wind load include the geographic location, elevation, degree of exposure, relationship to nearby structures, building height and size, direction of prevailing winds, velocity of prevailing winds and positive or negative pressures due to architectural design features (atriums, entrances, or other openings). All of these factors are taken into account when the lateral loads on the facades are calculated. It is often necessary to examine more than one wind load case. For this course, it will be assumed that wind loads, as well as the pressure they develop upon wall and roof elements, are static and uniform. They actually not only pound a structure with a constantly oscillating force, but also increase as a building increases in height. The loading of a tower can be very roughly approximated by an evenly distributed load. It is a vertical cantilever. The applet below allows you to investigate the variables which influence the structural behaviour of a tall, thin tower. It does not represent actual methods of calculating the total wind force on a tall building. It is intended to demonstrate the interaction between the variables of the equations which govern the structural behaviour.

Earthquake Loads

Earthquake loads are another lateral live load. They are very complex, uncertain, and potentially more damaging than wind loads. It is quite fortunate that they do not occur frequently. The earthquake creates ground movements that can be categorised as a "shake," "rattle," and a "roll." Every structure in an earthquake zone must be able to withstand all three of these loadings of different intensities. Although the ground under a structure may shift in any direction, only the horizontal components of this movement are usually considered critical in a structural analysis. It is assumed that a load-bearing structure which supports properly calculated design loads for vertical dead and live loads are adequate for the vertical component of the earthquake. The "static equivalent load" method is used to design most small and moderate-sized buildings.

The lateral load resisting systems for earthquake loads are similar to those for wind loads. Both are designed as if they are horizontally applied to the structural system. The wind load is considered to be more of a constant force while the earthquake load is almost instantaneous. The wind load is an external force, the magnitude of which depends upon the height of the building, the velocity of the wind and the amount of surface area that the wind "attacks." The magnitude earthquake load depends up the mass of the structure, the stiffness of the structural system and the acceleration of the surface of the earch. It can be seen that the application of these two types of loads is very different. This movie is a representation of the movement of a free standing water tower in an earthquake. It can be seen that the as the ground moves, the initial tendency is for the water tower to remain in place. The shifting of the ground is so rapid that the tower cannot "keep up."

After a moment, the tower moves to catch up with the movement of the ground. The movement is actually an acceleration. From Newtonian Physics, it is know that an applied force=mass x acceleration. Thus, the force which is applied to the water tower depends upon the mass of the tower and the acceleration of the earth's surface. The force in this last diagram may be thought of as the "equivalent static load" for which the structure would be designed.

This idealised situation demonstrates a concept; it requires modification for actual buildings. These modifications account for building location, importance, soil type, and type of construction. This movement can also be seen in the following movie of lateral earth movement. Note how the mass slowly reacts to the movement of the earth. Eventually, the bending strength of the stem of the tower would be exceeded and it will fail.

It remains very difficult to imagine the destruction which can be wrought by an earthquake. The lessons learned from the Los Angeles Earthquake of 1994 helped structural designers change design strategies. The following illustrations are snapshots from LA:

Fluid and Earth Pressure Loads

Liquids produce horizontal loads in many structures. The horizontal pressure of a liquid increases linearly with depth and is proportional to the density of the liquid. This is similar for earth pressures. These last are a bit more complex in that the load due to earth pressure varies with its depth, any surcharge, the type of soil and its moisture content. The design live load for this soil pressure must not be less than that which would be caused by a fluid weighing 30 pcf.

Load Distribution

Gravity Load Distribution

Architects must consider not only what loading conditions might effect a structure, but also how those loads are transferred from their point of origin through the structure to the ground. Architectural design decisions concerning the structural framing system will dictate this flow of forces. The most direct path from the point of loading to the ground will create the most efficient structure. The continuity of this line of transfer is essential. However, this direct path is not always possible. As a matter of fact, due to the nature of live loads this path is constantly changing!

The load at the midspan of the joist of the frame is illustrated as a large arrow pointing down. This load is split into two parts, with a proportional part going to each end of the beam. This in turn becomes a point load at the midspan of the smaller beams. They transfer the load to the earth. The load on each column in this case is only one quarter of the initial load. This is one method of distributing a load.

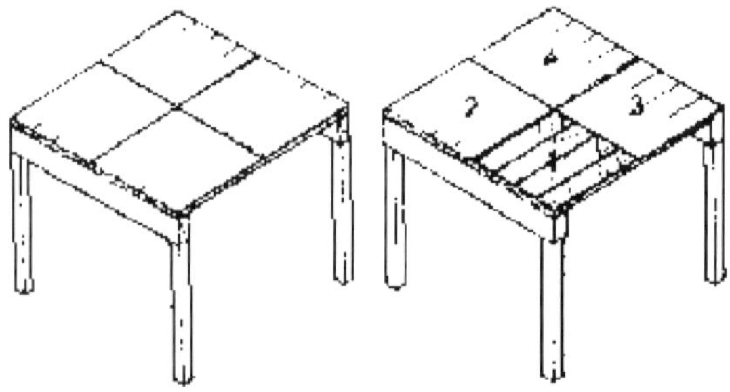

The second drawing illustrates a wooden floor of thick planks that span between two beams. Each of the four squares of the flat surfaces is the tributary area of the beam below it. A tributary area of a structural element (such as a joist, beam, column, or wall) is the area that contributes to the loading of that specific element. In this case, the plans transfer their loads and pass them on to the two beams. These beams can only transfer their loads equally to the two columns which support them. Thus, the tributary area can be drawn by determining the supporting characteristics of the horizontally spanning members.

The loads are gathered by each structural element and passed on to that element's supporting elements. These supports in turn pass their loads to the next supporting element until the original load has been transferred all the way to the earth.

The load distribution pattern and the explicit summation of the loads can have a direct effect upon the size of the elements. Load distribution often causes unequal loading of the vertical supporting members. This may or may not be indicated by the designed form of that element.

Wind Load Distribution

The essence of wind load distribution through a building from one building component to another, and how the building resists this load can be investigated by assuming a uniformly distributed wind load acting on any one face of the building at any one time. Normally a wind load design requires a separate analysis of wind from two perpendicular directions, such as wind from the north or south and then from the east or west. Consider a one story, flat roofed, rectangular, wood framed building without a parapet. The wall receiving the direct pressure from the wind distributes the top half of its horizontal wind load to the roof and the bottom half to the foundation; it acts as a vertical beam with the roof and foundation acting as simple horizontal supports. The part of the load going to the foundation is distributed from the foundation to the ground. The portion of the load going to the roof tends to cause the roof to move laterally; this lateral movement is resisted by the end walls. The movement of the end walls is prevented by their connection to the foundation.

Since winds of 80 to 110 mph will create pressure on the side walls of buildings in the range of 15 to 30 psf, design forces of the wind will be considered between 20 or 25 psf (static load) in this course.

Resolving Distributed Loads

Many structural members are subjected to distributed loads rather than concentrated loads. However, when solving for their reactions, it is often convenient to replace the distributed loads with an equivalent concentrated load(s).

The most common type of distributed load is the uniform load, in which a load of constant magnitude is applied along a length of a beam or an area of a surface. The resultant of these distributed loads has the same magnitude as the area of the distributed load and acts through the centre of gravity, or the centroid, of the load area. This point is located at the intersection of the diagonals of a rectangular, or uniform, load area.

Not all distributed loads are of a constant value. Some examples of loading conditions that may cause triangular loads are:

- the load on a lintel in a masonry wall due to the arching of the masonry;
- the load on a vertical surface below grade due to soil or water pressure;
- the load on the vertical surface of a structure due to wind loading.

These triangular loads can be replaced by a single concentrated load which acts through the centroid of the distributed load and acting in the same direction. The centroid of a triangular load area is found at the intersection of its medians which always equals 1/3 of the height above any of its bases.

Triangular loads can be caused by either soil or water pressure. Note that it constantly increases with depth! The magnitude of the greatest load that is applied is simply the unit weight of the material multiplied by the deepest point in question.

There are cases where a uniformly distributed load combines with a triangularly distributed load to create a trapezoidal load. This loading condition is most easily solved by separating the triangular and uniform loads and calculating separate resultants for each. These resultants can then be treated as a series of concentrated loads or transformed into a single resultant for the system.

Strength of Materials

A structural system is not only effected by external conditions, but also by the properties and behaviour of the materials which comprise it. These also determine the nature of the system's reaction(s) to external forces. The study of Strength of Materials is concerned specifically with the following issues:

1. The internal forces of a member caused by the external forces acting on that member or system.
2. The changes in dimension of a member caused by these forces.
3. The physical properties of the material in the member.

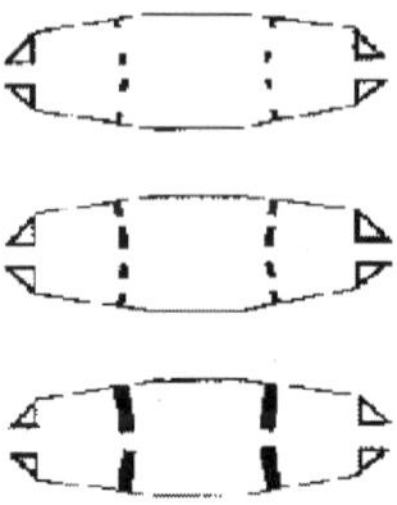

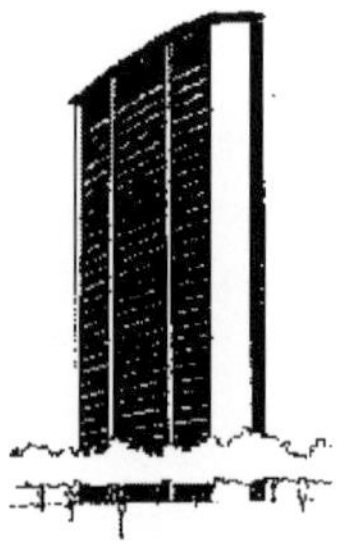

Statics is the study of the behaviour of rigid bodies at rest as they are acted upon by external forces. Although most of these bodies are not absolutely rigid, the assumption of rigidity is valid for the purpose of determining the reactions of the system. Actually, every material will deform under a load. Even a concrete slab deforms microscopically when a person walks on it. Some deformations in a structure can be detrimental to the overall system's performance, while others might only be an issue of comfort. The recognition of the relative importance of these deformations will be an important part of the study of structures. External loads on a structural system create resisting forces within all of the members that form the load path from the point of the application of the load to the ground beneath the foundation. This internal resistance exists within every member and joint included in the load path and are known simply as the internal forces acting on a member.

Some of these forces have already been examined: the connection between a beam and its support, and the connection of a two-force member to a three-force member in a pin-connected frame. The internal forces within a beam were demonstrated by cutting a beam and drawing a FBD. These internal forces were required at the cut section to put the beam back into equilibrium. The forces and moments that were examined were applied externally to the end of that cut section; they were exactly equal to the internal forces and moments. The distribution of the internal forces on the cross-sectional area of a member may or may not be uniformly distributed; it is dependent on the loading condition, the type of member, and how it is supported.

Stress and Stress Prisms

The internal forces of each member of a structural system are distributed in a particular way across that member's cross-sectional area. Stress is a measure of the intensity of this force on a single unit of area. Common units of measure of stress are KN/m^2, N/mm^2, $\#/in^2$, $\#/ft^2$, or $kips/in^2$. The distribution of the stress may be constant across a cross-sectional area, or it may be variable. It can vary due to the loading conditions, the material or geometry of the structural member.

Structural members are subjected to a variety of types of stress. They can be divided into primary and secondary stresses. Primary stresses are Compression, Tension, Torsion and Shear. Secondary stresses are the result of deformations of the structural system, or member, which cause additional loads within each member to occur (For example, a beam expanding in the summer sun pushing on a fixed column). These stresses will be discussed in more detail in a later lecture. Initially, the focus will be on simple tension and compression stresses for members loaded symmetrically along an axis, which causes uniformly distributed internal stress.

The Pirelli Office Block, designed by Ponti and engineered by Nervi, is a wonderful example of a structure in which the columns express not only the relationship between the moment and the building, but of the relationship between direct stress and size as well. The stress in the column was intended to remain as constant as possible.

Stress-Strain Curves

The relationship between the stress and strain that a material displays is known as a Stress-Strain curve. It is unique for each material and is found by recording the amount of deformation (strain) at distinct intervals of tensile or compressive loading. These curves reveal many of the properties of a material (including data to establish the Modulus of Elasticity, E). What does a comparison of the curves for mild steel, cast iron and concrete illustrate about their respective properties?

It can be seen that the concrete curve is almost a straight line. There is an abrupt end to the curve. This, and the fact that it is a very steep line, indicate that it is a brittle material. The curve for cast iron has a slight curve to it. It is also a brittle material. Both of these materials will fail with little warning once their limits are surpassed. Notice that the curve for mild steel seems to have a long gently curving "tail". This indicates a behaviour that is distinctly different than either concrete or cast iron. The graph shows that after a certain point mild steel will continue to strain (in the case of tension, to stretch) as the stress (the loading) remains more or less constant. The steel will actually stretch like taffy. This is a material property which indicates a high ductility. There are a number of significant points on a stress-strain curve that help one understand and predict the way every building material will behave. An example plot of a test on two grades of steel is illustrated above. If one begins at the origin and follows the graph a number of points are indicated. Point A is known as the proportional limit. Up to this point the relationship between stress and strain is exactly proportional. The number which describes the relationship between the two is the Modulus of Elasticity. This is discussed in more detail in the next lecture.

Strain increases faster than stress at all points on the curve beyond point A. Up to this point, any steel specimens that is loaded and unloaded would return to its original length. This is known as elastic behaviour. Point B is the point after which any continued stress results in permanent, or inelastic, deformation. Thus, point B is known as the elastic limit. Since the stress resistance of the material decreases after the peak of the curve, this is also known as the yield point. The line between points C and D indicates the behaviour of the steel specimen if it experienced continued loading to stress indicated as point C. Notice that the dashed line is parallel to the elastic zone of the curve (between the origin and point A). When the specimen is unloaded the magnitude of the inelastic deformation would be determined (in this case 0.0725 inches /inch). If the same specimen was to be loaded again, the stress-strain plot would climb back up the line from D to C and continue along the initial curve. Point E indicates the location of the value of the ultimate stress. Note that this is quite different from the yield stress. The yield stress and ultimate stress are the two values that are most often used to determine the allowable loads for building materials and should never be confused.

A material is considered to have completely failed once it reaches the ultimate stress. The point of rupture, or the actual tearing of the material, does not occur until point F. It is interesting to note the curve that indicates the actual stress experienced by the specimen. This curve is different from the apparent stress since the cross sectional area is actually decreasing. There is quite a bit to be learned from both the study of the ideal and actual behaviour of all building materials. Changes in that body of knowledge have had large impacts on the way in which building structures are designed.

The earliest methods of design limited the stresses that a structure would be "allowed" to experience. Thus, the method of design was known as the Allowable Stress Method. Recognition of the additional strength potential of most materials resulted in the Ultimate Stress Method of design. Contemporary thought centres on the limitation of the various service conditions of the structure at hand. This is known as the Limit States Design method. In the end, it is the authors opinion that the actual method of design is less important than the legal bodies would like us to believe. Human factors in the construction process SHOULD prevent a good designer from pushing too hard against the envelope of safety.

Elasticity and Deformation

An important property of many structural materials is their ability to regain their original shape after a load is removed. These materials are called *elastic*. Steel, glass and rubber are elastic; putty or modelling clay are not elastic. Each of these materials is elastic to varying degrees; steel and glass are both more elastic than rubber. The degree of elasticity, or "stiffness" of a material is called its Modulus of Elasticity (E). Given the modulus of elasticity, possible deformations can be calculated for any material and loading.

All materials are assumed to be elastic except when indicated otherwise in this course. Robert Hooke (1635 - 1703), a great English scientist, experimented with springs, clocks and watches. During his investigation of the spring he discovered that in elastic materials, stress and strain are proportional. He first presented this in a lecture in 1678 and it is known today simply as Hooke's Law.

Hooke's Law applies as long as the material stress does not pass a certain point known as its proportional limit. This is the point at which the physical properties of the material actually change. Any time an elastic material is loaded between zero and the proportional limit, the stress and strain are directly proportional and if the load is released the material will regain its initial dimensions. If the stress is doubled the strain is doubled; if the stress is tripled the deformation is three times as great, etc.

The English physician and physicist Thomas Young (1773 - 1829) noted that if stress is proportional to strain, then for any given material, stress divided by strain would be a constant. This constant is known today as Young's Modulus or the Modulus of Elasticity.

The Modulus of Elasticity is represented by E = Stress / Strain.

This relationship is found as the slope of the curve of the stress-strain curve from initial loading to the proportional limit. A higher value of the modulus indicates a more brittle material (i.e. glass, ceramics). A very low value represents a ductile material (i.e. rubber).

The values of the modulus of elasticity for structural materials have been determined by tests and are readily available in references such as the AISC manual. Some of the more common values are:

- Steel: E= 29,000 KSI (sometimes rounded to 30,000 KSI)
- Aluminum: E= 10,000 KSI
- Wood: E= 1,000 - 2,000 KSI (usual range)
- Concrete E= 2,500 - 5,000 KSI (usual range).

It is often necessary to be able to determine the deformation of a structural member once the loads and physical properties of the structural member are known. This is simply derived and is developed from the stress/strain relationships that have already been established.

Secondary Loads

Some types of loads on buildings are due to deformation. The deformation may be due to earth sliding or settling unevenly under the footings. This type of load can be avoided by properly designed footings on stable soil. The deformation of a building may be caused by temperature change. To avoid excessive stresses, construction must provide for a certain amount of thermal expansion and contraction.

Built-up Sections

It is often advantageous to combine a number of smaller members in order to create a beam or column of greater strength. The moment of inertia of such a built-up section is found by adding the moments of inertia of the component parts. This can be done< B> if and only if the moments of inertia of each component area are taken about a common axis and if and only if the resulting section acts as one unit.

Under No Other Condition Can They Be Added!

Two examples of built-up sections are seen below. In each case the components of the whole have a common axis and act as one unit.

Transfer Formula

There are many built-up sections in which the component parts are not symmetrically distributed about the centroidal axis. The easiest way to determine the moment of inertia of such a section is to find the moment of inertia of the component parts about their own centroidal axis and then apply the transfer formula. The transfer formula transfers the moment of inertia of a section or area from its own centroidal axis to another parallel axis. It is known from calculus to be:

$$I_x = I_c + (A)d^2$$

Where:

I_x = moment of inertia about axis x-x (in^4)

I_c = moment of inertia about the centroidal axis c-c parallel to x-x (in^4)

A = area of the section (in^2)

d = perpendicular distance between the parallel axes x-x and c-c (in)

Finding the moment of inertia of an asymmetric built-up cross-section is simplified to the procedure shown diagrammatically below:

Columns

The column is one of the most familiar architectonic icons. Each of us can picture this very important space definer in our mind's eye. It can delineate, punctuate, or separate spaces. A quick glance out most windows allows a view of at least one of them. Columns are vertical load-bearing elements which are normally loaded in compression. Time has brought many manifestations of this simple element; thick, thin, long, short, spindle shaped, squat, etc. And they have been made out of any material that has a minimal compressive strength.

The transition between the column and its loading member, usually a beam of some sort, has often been of great interest to builders. The structural/functional problem is quite simple - the forces collected by the beam must be transferred from the beam to the column - and yet the solutions demonstrated a great variation. The early Egyption, Greek and Roman civilizations turned this transition zone into an articulated flourish to draw attention to the capital. In almost every case it was broader at the top and tapered to the column shaft. This not only facilitated a smooth transition for the forces, but also eased the construction and even increased the stability of the buildings.

A column either crushes (a strength failure) or it buckles (a stability failure). Both modes of failure must be considered for every column. The exact mode of failure is greatly dependent upon the way in which the column's cross-sectional area is distributed with respect to its centroid. The following simple concept must be satisfied at all times:

Stress Due to Loading < Resistance Potential of the Column

This states that the stress within the column due to all of the applied loads must be less than the allowable stress of the material. This is a logical statement that is the essence of the structural analysis of a column. The actual failure mechanism could be due to a combination of two or more loads. These combinations must be carefully considered.

Column Failure Modes

Crushing: Relatively short columns are more apt to fail by the material crushing. Every building material material can withstand a distinct amount of compressive stress before it crushes. This value has been determined by laboratory tests and is known as the compressive strength of a material. This strength is dependent upon the internal structure of the material and/or its components. Steel has a very homogeneous, finely crystalled internal structure and has a relatively high compressive strength. Hard woods are fine-grained and have a higher compressive strength than soft-woods. Wood is unique in that it has two compressive strengths; one when loaded parallel to the grain and another when loaded perpendicular to the grain. Why is that? When a wood column crushes the fibres of the wood actually split apart. In every case, crushing is a strength failure and does not depend upon the shape of the section.

Buckling: Relatively slender columns are more apt to fail by buckling. A column is slender when it has a "small" cross-section compared to its effective length. Small is placed in quotes due to the fact that the important information about the cross section is both the actual size and more importantly, the shape of the cross-section. This is then compared to the effective length to determine whether or not the column is slender. If it is, this means that the column will probably fail in bending! As a column is loaded, it is likely to bend about the weak axis of the cross-section (the one with the lowest Moment of Inertia). A column buckles when it bends about an axis. This is a stability failure.

Radius of Gyration and Buckling

The radius of gyration (r) describes the way in which the area of a cross-section is distributed around its centroidal axis. If the area is concentrated far from the centroidal axis it will have a greater value of r and a greater resistance to buckling. A cross-section can have more than one radius of gyration and most sections have at least two. If this is the case, the section tends to buckle around the axis with the smallest value. The radius of gyration is defined as:

r = sqrt (I/A)

where

r = radius of gyration

I = moment of inertia

A = area of the cross section

All things being equal, a circular pipe is the most efficient column section to resist buckling. This is because it has an equal radius of gyration in all directions and it has the its area distributed as far away as possible from the centroid.

The steel columns shown below all have areas of 3-1/8 in^2. The safe loads for an 8 ft length are shown. The only difference between them is the way in which the cross-sectional area is distributed about the centroid.

Buckling

Buckling is very similar to bending. Thus, the shape of the cross-section is very important. The shape of the column also effects the way in which it will buckle. Imagine for a moment a single sheet of paper (A4 or 8.5 x 11). If one would try to simply stand it on edge it would be impossible unless the paper was folded. This simple act of folding the paper actually increases the cross-sectional moment of Inertia and thus the stiffness of the newly formed column.

The stiffness of another paper column could be further increased by taking the paper and taping the long edges together to create a tube. Now, the paper will be very stiff since the material of the paper is distributed evenly as far away from the neutral axis as possible.

The load at which a column will begin to buckle is known as the Critical Buckling Load (or critical load). A number of qualities of the column must be known in order to determine this value. The Swiss mathematician Leonard Euler (1707 - 1783) derived a formula in 1744 (known as the Euler Buckling Formula) to determine the load at which a perfect column will buckle.

It was a very important step in the history of technology and remains important for column design today. The equation is only accurate for columns which approach the perfect conditions for which he derived the equation.

$N_{cr} = \frac{1}{4}^2 E\ I\ /\ (l_k^2)$

In which the terms are defined as follows:

N_{cr} = Critical Buckling Load

E = Modulus of Elasticity

I = Moment of Inertia

l_k = Effective Buckling Length

This equation can be modified by dividing both sides by the area of the column so that the stress at which the column will buckle can be determined:

$f_{cr} = sigma_{cr} = N_{cr}/A = \frac{1}{4}^2\ E\ I\ /A(l_k^2)$

now, knowing that $r^2 = I/A$, this equation becomes:

$= \frac{1}{4}\ ^2 E\ /(l_k/r)^2$

Lambda, or the slenderness ratio is a value with which one can gage the relative resistance of a column cross-section to buckling. Or, stated otherwise, the relative ease in which a column WILL buckle. It is defined as

$lambda = l_k\ /\ r$

Where l_k is the buckling length and r is the radius of gyration.

Thus, the critical buckling stress can be expressed as,

f_{cr} = ¼^2 E / (lambda)^2

The buckling length of a column depends on its physical length and its end conditions. Euler discovered that if a column is hinged at both ends it will buckle in the form of a sine curve with the inflection points at the hinges. This would be the case in which the buckling length of a column is identical to its length. This is not the case if the ends of the column are both fixed. The determination of the buckling lengths for various column end conditions and frames is given below:

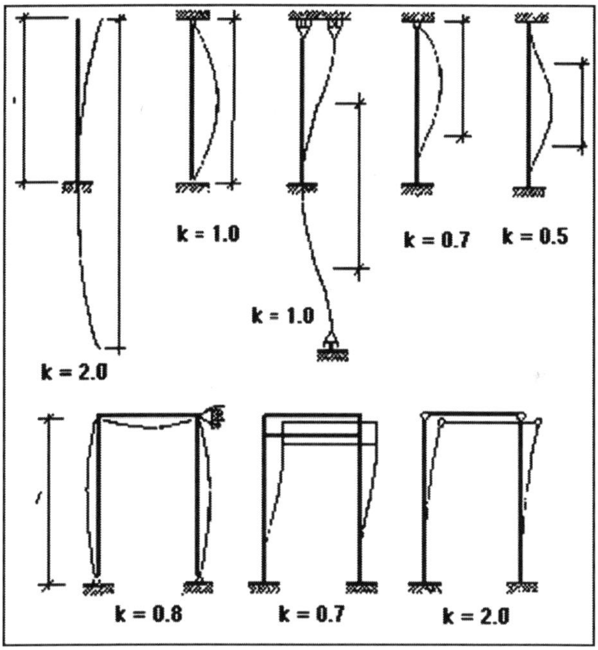

The magnitude of the internal forces is important to know in order to size a column. A small amount of tensile stress has little effect on a wood or steel column, but could cause problems in a concrete or masonry column.

Beams

A beam is a structural member which carries loads. These loads are most often perpendicular to its longitudinal axis, but they can be of any geometry. A beam supporting any load develops internal stresses to resist applied loads. These internal stresses are bending stresses, shearing stresses, and normal stresses. Beam types are determined by method of support, not by method of loading. Below are three types of beams that will be investigated in this course:

1. Simple 

2. Cantilever 

3. Continuous 

The first two types are statically determinate, meaning that the reactions, shears and moments can be found by the laws of statics alone. Continuous beams are statically indeterminate. The internal forces of these beams cannot be found using the laws of statics alone. Early structures were designed to be statically determinate because simple analytical methods for the accurate structural analysis of indeterminate structures were not developed until the first part of this century. A number of formulas have been derived to simplify analysis of indeterminate beams.

The three basic beam types can be combined to create larger beam systems. These complex systems can *inevitably* be distilled to the simple beam types for analysis. The beams shown immediately below are combinations of the first two beam types; these systems are all statically determinate.

Concentrated

Either a force or a moment can be applied as a concentrated load. Both are applied at a single point along the axis of a beam. These loads are shown as a "jump" in the shear or moment diagrams. The point of application for such a load is indicated in the diagram above. Note that this is NOT a hinge! It is a point of application. This could be point at which a railing is attached to a bridge, or a lamppost on the same.

Distributed

Distributed loads can be uniformly or non-uniformly distributed. Both types are commonly found on all kinds of structures. Distributed loads are shown as an angle or curve in the shear or moment diagram. A uniformly distributed load can evolve into a n on-uniformly distributed load (snow melting to ice at the edge of a roof), but are normally assumed to act as given. These loads are often replaced by a singular resultant force in order to simplify the structural analysis.

Beam Failure Modes

Structures fail in many ways; two categories of which have been described: stability failures and strength failures. Stability failures usually relate to structural systems, whereas strength failures relate to the members comprising a structure. Of the many ways which a beam can fail, three will be discussed in this course: bending, vertical shear, and horizontal shear.

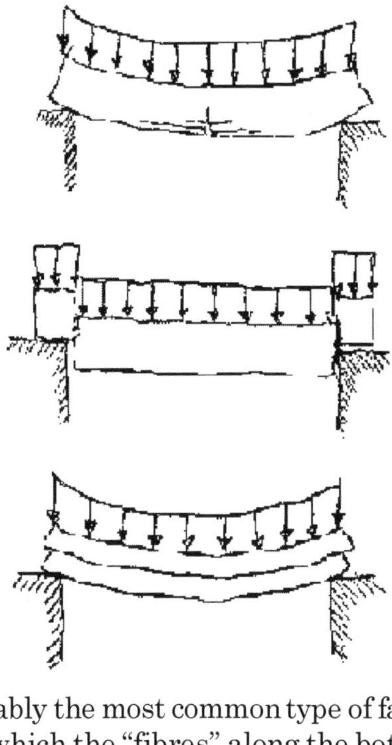

Bending is probably the most common type of failure. It is illustrated by the top figure in which the “fibres” along the bottom face of the beam are torn and those along the top face of the beam are crushed. Vertical Shear is an idealised mode of failure. There is a tendency for a short beam to fail in this manner. This is the way in which a pair of “shears” or scissors cuts a piece of paper. Horizontal Shear is the tendency for a material to separate parallel to the neutral axis as its “internal layers” try to slide past each other. It is a frequent mode of failure which should not be confused with checking in wood beams. Our discussion during the next few days will be limited to the stresses that are related to these modes of failure. Bending and horizontal shear are the types of failure most often encountered. Vertical shear is useful to us primarily because of what it reveals about horizontal shear. Excessive deflection can also be a cause of failure.

Internal Forces

Internal forces are generated within loaded structural elements. These forces are generated within every type of element; if they were not developed, the structure would fail. These are known as Shear, Moment, and Normal Forces. The normal force is found in columns and beams with an axial load. Shear and moment are found in beams and frames. Most of the elements that will be analysed in this course will be beams, including joists, purloins, girders, decking, planking, etc. Shear and moment are essential for developing an understanding of how structures behave. Considerable time will be spent on this subject because of its importance now and in the future. The capacity to draw shear and moment diagrams for any designed structure will add to the repertoire of the designer. All of the great Modern architects had this basic skill. In order to quantify the internal forces of any part of a beam a FBD must be drawn for *only* that specific part of the beam. If the entire structure is in equilibrium, this part, when isolated from the remainder as a Free Body, must also be in equilibrium. The action of the external forces on that part must be resisted by the internal forces acting at the cut section(s).

External Moment = Internal Moment or External Moment - Internal Moment = 0

Internal resisting forces and moments caused by the loading on a beam usually change along the length of the beam. Let us again examine the internal forces of a weightless beam by means of a FBD:

A beam carrying a load is clearly in equilibrium. If we cut the beam and apply the forces on the cut section which existed internally before the cut, the beam will still be in equilibrium. We can find these forces by applying equations of equilibrium on the ends of the cut sections (as in any FBDs). The solutions to the equations of equilibrium are the internal normal (N), shear (V) and moment (M) forces at that section. This is then applied to each and every point along the beam.

Shear (V) and Shear Diagrams

Shear (V) is the tendency for one part of a beam to slide past another part. The magnitude of the shear at any section is equal to the algebraic sum of loads and reactions acting perpendicular to that section.

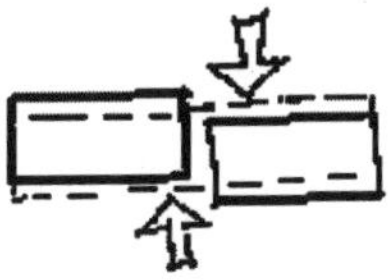

Sign Convention

If the tendency of the section to the left of the cut is to move upward, the shear is positive; if it has a tendency to move down, it is negative. For simplicity in obtaining the correct sign, one may say that it is equal to the algebraic sum of loads and reactions to the left of the cut. Note that the internal shear on the end of the FBD acts in the opposite direction of the algebraic sum of the loads and reactions in order to balance these forces (SV = 0).

A shear diagram is a graphic representation of the shear at every point along the length of a member.

To plot a shear diagram on a beam, the shear must be calculated at each point along the length of the beam. One way is to simply plot the shear as the algebraic sum of the loads and reactions acting perpendicular to the beam at the left side of each increment along the length of the beam. Positive values are shown above and negative values below a reference axis.

How to Draw a Shear Diagram

Start at the left end and plot the external shear values with regard to the following:

- The shear diagram is the graphic representation of the shear force at successive points along the beam. Forces acting upward are assumed positive and downward forces negative.
- The shear force (V) at any point is equal to the algebraic sum of the external loads and reactions, perpendicular to the beam, to the left of that point.
- Since the entire beam must be in equilibrium (sum of V = 0), the shear diagram must close to zero at the right end.
- Consider the loading for increments along the length of the beam in order to determine the shape of the curve.

If there is no change in the load along the incremental length under consideration, the shear curve is a straight horizontal line (or a curve of zero slope). The slope at any point is defined as the tangent to the curve at that point.

If a load exists, but does not change in magnitude over successive increments (uniformly distributed), the slope of the shear curve is constant and non-horizontal.

If a load exists, and increases in magnitude over successive increments, the slope of the shear curve is positive (approaches the vertical); if the magnitude decreases, the slope of the shear curve is negative (approaches the horizontal).

- Abrupt changes in loading cause abrupt changes in the slope of the shear curve. Concentrated loads produce vertical lines (a jump) in the shear curve.

Moment (M) and Moment Diagrams

The Moment (M) within a beam is a representation of the magnitude of the internal couple found within the beam at any given point. A moment diagram is a graphic representation of the moment at every point along the length of the member. To plot a moment diagram of a beam, one must determine the magnitude of the moment at each point along the length of the beam and then determine whether it is a positive or negative moment. Positive values are shown above and negative values below a reference axis. Sign convention is probably the most troublesome part of moment diagrams. In North America, if the moment tends to cause the beam to curve upward it is positive; if the moment tends to cause it to curve downward it is negative. This is the opposite of the conventions for most of the rest of the world. The sign of the moment diagram is not necessarily the same sign of the moment found in the FBD or in the moment equation when determining its magnitude. A simply supported beam with gravity load always has a *positive* moment. A cantilever with gravity load always has a *negative* moment. A continuous beam with an overhang subject to a gravity load experiences both *positive and negative* moments.

How to Draw a Moment Diagram

Start at the left end and plot the bending moment values due to external loads and reactions with regard to the following:

- The moment diagram is the graphic representation of the magnitude of the bending moment at successive points along the beam.

- The bending moment for the moment diagram (M) at any point equals the sum of moments of the forces to the left about that point.
- Since the entire beam is in equilibrium (sum of M= 0), the bending moment diagram must close to zero at the right side.
- Consider the values of successive incremental ordinates along the length of the beam in order to determine the shape of the curve:

If the magnitudes of successive shear ordinates are constant, the moment curve has a constant slope at that increment. If the magnitudes of successive shear ordinates increase, the slope of the moment curve is positive (it approaches the vertical). If the magnitudes of successive shear ordinates decreases, the slope of the moment curve is negative (it approaches the horizontal).

- Abrupt changes in the shear diagram will produce abrupt changes in the slope of the moment curve. Concentrated moments produce vertical lines in the moment curve.

Plotting V, M and Deflection

A comprehensive procedure for the construction of shear, moment, and deflection diagrams for statically determinate beams is given below. Each and every one of these diagrams are plotted from left to the right and positive values are shown above, and negative values below, a reference axis. The procedure is as follows:

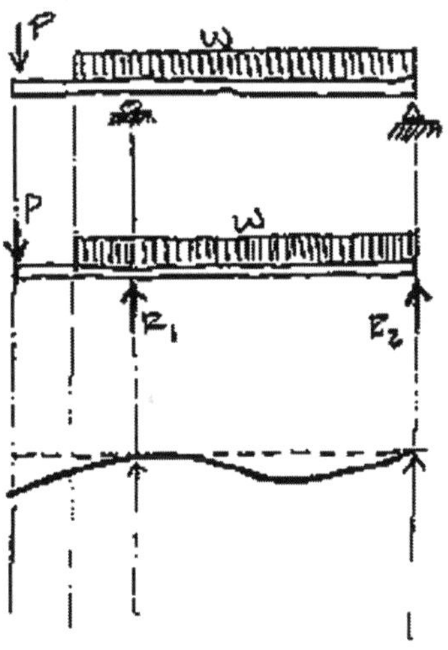

1. Draw a FBD of the entire structure and find the reactions of the beam.

2. Based on intuition and an educated guess, sketch an approximate deflection curve of the structure. This will help to determine the location of inflection points and thus help to construct the diagrams.
3. Draw a SHEAR DIAGRAM. Start at the left end and plot the external shear values with regard to the following:
   - The shear diagram is the graphic representation of the shear force at successive points along the beam. Upward acting forces are assumed positive and downward forces negative.
   - The shear force (V) at any point is equal to the algebraic sum of the external loads and reactions to the left of that point.
   - Since the entire beam must be in equilibrium (the sum of V = 0), the shear diagram must close to zero at the right end.
   - Consider the loading for increments along the beam's length in order to determine the shape of the curve.

If there is no change in the load along the incremental length under consideration, the shear curve is a straight horizontal line (or a curve of zero slope). The slope at any point is defined as the tangent to the curve at that point.

If a load exists, but does not change in magnitude over successive increments (uniformly distributed), the slope of the shear curve is constant and non-horizontal.

If a load exists, and increases in magnitude over successive increments, the slope of the shear curve is positive (approaches the vertical); if the magnitude decreases, the slope of the shear curve is negative (approaches the horizontal).

   - Abrupt changes in loading cause abrupt changes in the slope of the shear curve. Concentrated loads produce vertical lines (a jump) in the shear curve.

4. Draw a MOMENT DIAGRAM. Start at the left end and plot the bending moment values due to the external loads and the reactions with regard to the following:
   - The moment diagram is the graphical representation of the magnitude of the bending moment at successive points along the beam.
   - The bending moment for the moment diagram (M) at any point equals the sum of moments of the forces to the left about that point.

- Since the entire beam is in equilibrium (Sum of M= 0), the bending moment diagram must close to zero at the right side.
- Consider the values of successive incremental ordinates along the length of the beam in order to determine the shape of the curve:

If the magnitudes of successive shear ordinates are constant, the moment curve has a constant slope at that increment. If the magnitudes of successive shear ordinates increase, the slope of the moment curve is positive (it approaches the vertical). If the magnitudes of successive shear ordinates decreases, the slope of the moment curve is negative (it approaches the horizontal).

- Abrupt changes in the shear diagram will produce changes in the shape of the moment curve. Concentrated moments produce vertical lines in the moment curve.

Deflection Diagram

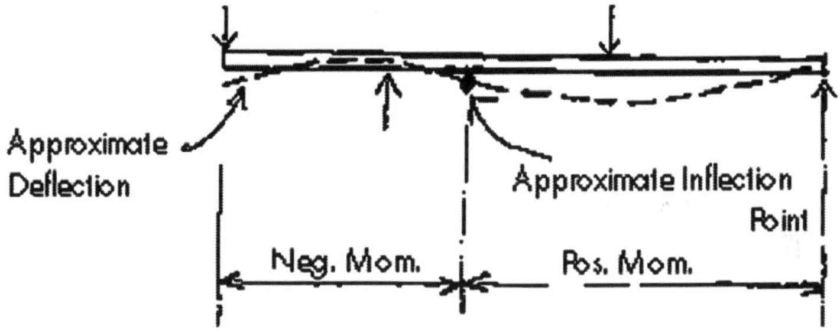

There is no set manner in which the deflection diagram should be drawn. There are, however, a few rules which can be followed:

- points of zero moment are inflection points (change in curvature direction) of the deflection curve.
- the maximum points of the moment diagram are the points at which the deflection diagram slope is zero (where the tangent of the curve is horizontal).

There are a few relationships (based on calculus) between these three diagrams that are helpful to construct them:

- The slope of the moment curve at any point is equal to the shear force at the same point.
- Points of zero shear are peaks of the moment curve, points of maximum positive or maximum negative moment.

- The change in moment value between any two points is equal to the area of the shear diagram between the same two points (provided no other external moment is applied). This is known as the Shear Area Method.
- The shapes of the curves from shear to moment always progress in this order; straight horizontal line, straight sloped line, parabolic curve and cubic curve. Therefore, if a shear curve is a horizontal line, the moment curve is a sloped line; if th e shear curve is a sloped line the moment curve is a parabolic curved line; etc. This happens because of the mathematical relationship between loading, shear and moment diagrams.

Beam Stresses

Bending Stresses: A formula based on bending in a beam will be derived that can be used to:

- Select a beam that has adequate strength to carry a given load in bending.
- Determine the bending stresses in a given beam caused by a specific loading condition.
- Determine the greatest moment a given beam will resist (or the greatest load it will carry)

It will not be necessary to be able to derive the formula. However, it is essential to work through the derivation at least once in order to understand it and have confidence in how it is applied.

$$F = M/S \text{ or } S = M/F$$

In which,

S is the Section Modulus in in^3. This beam property can either be calculated or read from tables for most beams.

M is the moment in the beam in inch-pounds (or inch-kips). In the selection of a beam it is taken from the moment diagram.

F is the bending stress (psi or ksi) (Fb is usually used to denote the allowable bending stress; fb is used to indicate the actual bending stress).

Bending Theory

There are a number of assumptions that were made in order to develop the *Elastic Theory of Bending*. These are:

- The beam has a constant, prismatic cross-section and is constructed of a flexible, homogenous material that has the same Modulus of Elasticity in both tension and compression (shortens or elongates equally for same stress).

- The material is *linearly elastic*; the relationship between the stress and strain are directly proportional.
- The beam material is not stressed past its proportional limit.
- A plane section within the beam before bending remains a plane after bending.
- The neutral plane of a beam is a plane whose length is unchanged by the beam's deformation. This plane passes thought the centroid of the cross-section.

In order to visualize this, think of a black rubber beam with three lines drawn on its side. The dashed lines in the diagram below represent this beam, and the neutral axis and lines AB and CD are drawn parallel to their respective sides. Lines AB and C D are separated by some distance that is determined, but is not of consequence for the following discussion. These two lines were parallel before bending. As the beam bends, these lines remain perpendicular to the neutral axis.

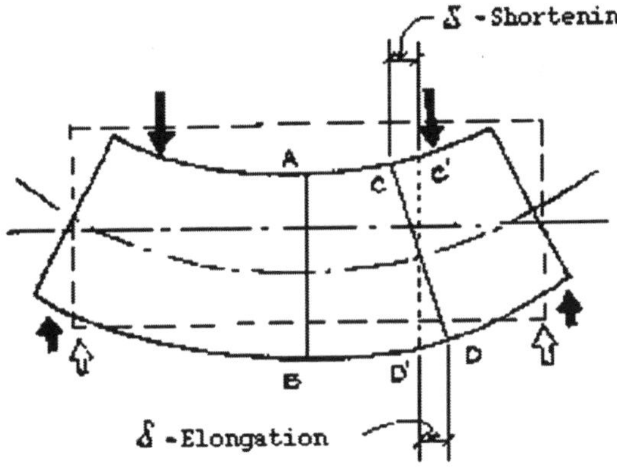

Thus, as the beam bends, developing a curve in the neutral plane that reflects the bending, the line CD does not remain parallel to line AD. There is a distinct shortening at the top face of the beam and elongation at the bottom face. CC' is equal to d (delta), or the shortening of the top fibre. Similarly, DD' is equal to (delta) or the elongation of the bottom fibre (tension). If these were measured, they would be found to be equal, but opposite. Knowing that the Modulus of Elasticity (E) describes a linear relationship between the strain and the amount of stress in a material and that this material is homogeneous and has a certain value for E, we can determine the stress. The magnitude of the strain at any point along C'D' can be found by using similar triangles. Therefore, the stress is also proportional to the distance from the neutral plane, as illustrated in the following diargram.

The loads and reactions acting on this beam segment cause a tendency for clockwise rotation (a clockwise moment). An internal moment counteracts this tendency so that the beam segment remains in equilibrium. Thus, the magnitude of the internal moment is exactly equal to the moment due to the external loads and reactions; but in the opposite direction. There is a state of equilibrium:

Internal Moment = External Moment

The illustration above shows the stress prisms developed due to the straining of the material. These prisms (compression and tension) can each be resolved into a single force that acts through the centroid of each prism. These two resulting forces (one is a compressive force (C) and the other a tensile force (T)) are found to have equal but opposite magnitudes. They are separated by a distance that is approximately 2/3 of the depth of the beam (because of their triangular geometry) and create an *internal couple(!)* which causes the internal moment resistance of the beam. The exact location of the resultant forces of the stress prisms depends upon the location of the centroid of the entire stress prism. The location of the centroid is in turn dependent upon the exact distribution of the stresses across the section which this is determined by the geometry of the cross-section.

General Case

If one accepts the assumptions given in the discussion of the bending theory above, the formula for determining the bending stress at any given fibre in a section can be derived as follows.

Given the beam as shown above loaded so that it has an internal bending moment. The stress at the extreme face (the distance C in this case) from the Neutral Axis (NA) is equal to Sigma or fb. The stress at any distance "y"; from the NA can be found using similar triangles. Thus,

$$fy = (fb/c)(y)$$

If "a" is a very small area at a distance "y" from the NA, then magnitude of the stress on area "a" equals (fb/c)(y). Remembering that stress is equal to a force distributed over an area, the force on the area "a" is equal to the stress times that area, or (fb/c)(y)(a). And, the moment of the force on area "a" about the NA is equal to (fb/c)(y)(a)(y).

Now, if all of the moments of all of the forces on all of the tiny "a" areas across the face of the beam are added together we arrive at an expression of the internal moment resistance of a beam;

$$M = \text{Sum } (f/c)ay^2 \text{ or } M = (f/c) \text{ Sum } ay^2$$

In this case, the value of (f/c) is a constant for any beam and Sum ay^2 is the second moment of the area, better known as the Moment of Inertia. Therefore, substituting items that we know, the equation becomes:

$$M = (fb/c)\ I$$

or rearranging

$$M = (I/c)\ fb$$

or

$$M = S*fb$$

(I/c) is known as the Section Modulus (S) and is also tabulated.

Shearing Stresses

For small members in shear, it is customary to assume shear is uniformly distributed over the entire area. However, this assumption is not permissible for beam cross sections. The vertical shearing stress at any point in a beam may be determined from the horizontal shearing stress at that cross-section.

- The infinitesimally small particle "A", whose dimensions are a, b, and 1, lies in a plane subject to shear.
- fs represents the vertical shearing stress and fv represents the horizontal shearing stress.
- fsb1 is the total shearing force on one vertical surface (force is equal to stress fs times the area b1); since the particle is in equilibrium an equal and opposite force must be acting on the other parallel face.
- Similarly, fva1 represents the total shearing force on each of the horizontal planes.
- The particle is in equilibrium, therefore the sum of the moments about any axis must equal zero. Take moments about axis "O":

This shows that at any point, the horizontal shearing stress is equal to the vertical shearing stress.

The general shear formula will find the shearing stress at any point in a beam. It is:

$$fv = V\ A'\ y\ /\ I\ b$$

V = the absolute value of vertical shear from the shear diagram at the point where shear is being investigated.

A' = the area above or below the horizontal plane being investigated.

y = the distance from the NA to the centroid of A'.

I = the moment of inertia of the entire section

b = the width of beam at the point where shear is being investigated.

Before deriving the formula, we will look at an application. If we wanted to find the maximum stress in the glue-line for this 2 x 10 composite beam:

- First draw the V diagram.
- Next, determine the geometry and associated properties of the beam and the appropriate value for y.
- Then plug into the equation to determine the stress at the glue-line.
- In this case, the stress would be 3000# (4")(2")(3") / (1/12)(2")(10^3)(2") or 216 psi.

Note: This is not a *bending* stress, it is a shearing stress.

10

Construction Surveying

Construction surveying (otherwise known as "layout" or "setting-out") is to stake out reference points and markers that will guide the construction of new structures such as roads or buildings. These markers are usually staked out according to a suitable coordinate system selected for the project.

History of Construction Surveying

- The nearly perfect squareness and north-south orientation of the Great Pyramid of Giza, built c. 2700 BC, affirm the Egyptians' command of surveying.
- A recent reassessment of Stonehenge (c. 2500 BC) suggests that the monument was set out by prehistoric surveyors using peg and rope geometry.
- In the sixth century BC geometric based techniques were used to construct the tunnel of Eupalinos on the island of Samos.

Elements of the Construction Survey

- Survey existing conditions of the future work site, including topography, existing buildings and infrastructure, and underground infrastructure whenever possible (for example, measuring invert elevations and diametres of sewers at manholes);
- Conduct an As-Built survey: a survey conducted at the end of the construction project to verify that the work authorised was completed to the specifications set on plans;
- Stake out reference points and markers that will guide the construction of new structures.

Coordinate Systems used in Construction

Land surveys and surveys of existing conditions are generally performed according to geodesic coordinates. However for the purposes of construction a more suitable coordinate system will often be used. During construction surveying, the surveyor will often have to convert from geodesic coordinates to the coordinate system used for that project.

Chainage or Station

In the case of roads or other linear infrastructure, a *chainage* will be established, often to correspond with the centre line of the road. During construction, structures would then be located in terms of *chainage, offset* and *elevation*. *Offset* is said to be "left" or "right" relative to someone standing on the *chainage line* who is looking in the direction of increasing *chainage*. Plans would often show *plan* views (viewed from above), *profile* views (a "transparent" section view collapsing all section views of the road parallel to the *chainage*) or *cross-section* views (a "true" section view perpendicular to the *chainage*). In a *plan* view, *chainage* generally increases from left to right, or from the bottom to the top of the plan. *Profiles* are shown with the chainage increasing from left to right, and *cross-sections* are shown as if the viewer is looking in the direction of increasing *chainage* (so that the "left" *offset* is to the *left* and the "right" *offset* is to the *right*). "Chainage" may also be referred to as "Station".

Building Grids

In the case of buildings, an arbitrary system of grids is often established so as to correspond to the rows of columns and the major load-bearing walls of the building. The grids may be identified alphabetically in one direction, and numerically in the other direction (as in a road map). The grids are usually but not necessarily perpendicular, and are often but not necessarily evenly spaced. Floors and basement levels are also numbered. Structures, equipment or architectural details may be located in reference to the floor and the nearest intersection of the arbitrary axes.

Other Coordinate Systems

In other types of construction projects, arbitrary "north-south" and "east-west" reference lines may be established, that do not necessarily correspond to true coordinates.

Equipment and Techniques used in Construction Surveying

Surveying equipment, such as levels and theodolites, are used for accurate measurement of angular deviation, horizontal, vertical and slope distances. With computerisation, electronic distance measurement (EDM), total stations, GPS surveying and laser scanning have supplemented (and to a large extent supplanted) the traditional optical instruments. The builder's level measures neither horizontal nor vertical angles. It simply combines a spirit level and telescope to allow the user to visually establish a line of sight along a level plane. When used together with a graduated staff it can be used to transfer elevations from one location to another.

An alternative method to transfer elevation is to use water in a transparent hose as the level of the water in the hose at opposite ends will be at the same elevation.

Equipment and Techniques used in Mining and Tunneling

Total stations are the primary survey instrument used in mining surveying.

Underground Mining

A total station is used to record the absolute location of the tunnel walls (stopes), ceilings (backs), and floors as the drifts of an underground mine are driven. The recorded data is then downloaded into a CAD program, and compared to the designed layout of the tunnel.

The survey party installs control stations at regular intervals. These are small steel plugs installed in pairs in holes drilled into walls or the back. For wall stations, two plugs are installed in opposite walls, forming a line perpendicular to the drift. For back stations, two plugs are installed in the back, forming a line parallel to the drift.

A set of plugs can be used to locate the total station set up in a drift or tunnel by processing measurements to the plugs by intersection and resection.

Professional Status of Construction Surveyors

Building Surveying emerged in the 1970's as a profession in the United Kingdom by a group of technically minded General Practice Surveyors. Building Surveying is a recognised profession within Britain and Australia.

In Australia in particular, due to risk mitigation/limitation factors the employment of surveyors at all levels of the construction industry is widespread. There are still many countries where it is not widely recognised as a profession. The Services that Building Surveyors undertake are broad but include:

- Construction design and building works
- Project Management and monitoring
- CDM Co-ordinator under the Construction (Design & Management) Regulations 2007
- Property Legislation adviser
- Insurance assessment and claims assistance
- Defect investigation and maintenance adviser
- Building Surveys and measured surveys
- Handling Planning applications
- Building Inspection to ensure compliance with building regulations
- Undertaking pre-acquisition surveys
- Negotiating dilapidations claims.

Building Surveyors also advise on many aspects of construction including:

- design
- maintenance
- repair
- refurbishment
- restoration.

Clients of a building surveyor can be the public sector, Local Authorities, Government Departments as well as private sector organisations and work closely with architects, planners, homeowners and tenants groups. Building Surveyors may also be called to act as an expert witness. It is usual for building surveyors to undertake an accredited degree qualification before undertaking structured training to become a member of a professional organisation. For Chartered Building Surveyors, these courses are accredited by the Royal Institution of Chartered Surveyors. Other professional organisations that have building surveyor members include CIOB, ABE, HKIS and RICS.

With the enlargement of the European community, the profession of the Chartered Building Surveyor is becoming more widely known in other European states, particularly France. Chartered Building Surveyors, where many English speaking people buy second homes.

Distinction from Land Surveyors

In the United States, Canada, the United Kingdom and most Commonwealth countries land surveying is considered to be a distinct profession. Land surveyors have their own professional associations and licencing requirements. The services of a licenced land surveyor are generally required for boundary surveys (to establish the boundaries of a parcel using its legal description) and subdivision plans (a plot or map based on a survey of a parcel of land, with boundary lines drawn inside the larger parcel to indicated the creation of new boundary lines and roads).

Transport Engineering

Transportation engineering is the application of technology and scientific principles to the planning, functional design, operation and management of facilities for any mode of transportation in order to provide for the safe, rapid, comfortable, convenient, economical, and environmentally compatible movement of people and goods (transport). It is a sub-discipline of civil engineering. Transportation engineering is a major component of the civil engineering discipline. The importance of transportation engineering within the civil engineering profession can be judged by the number of divisions in ASCE (American Society of Civil Engineers) that are directly related to transportation. There are six such divisions (Aerospace; Air Transportation; Highway; Pipeline; Waterway, Port, Coastal and Ocean; and Urban Transportation) representing one-third of the total 18 technical divisions within the ASCE (1987).

The planning aspects of transport engineering relate to urban planning, and involve technical forecasting decisions and political factors. Technical forecasting of passenger travel usually involves an urban transportation planning model, requiring the estimation of trip generation (how many trips for what purpose), trip distribution (destination choice, where is the traveller going), mode choice (what mode is being taken), and route assignment (which streets or routes are being used). More sophisticated forecasting can include other aspects of traveller decisions, including auto ownership, trip chaining (the decision to link individual trips together in a tour) and the choice of residential or business location (known as land use forecasting). Passenger trips are the focus of transport engineering because they often represent the peak of demand on any transportation system.

A review of descriptions of the scope of various committees indicates that while facility planning and design continue to be the core of the transportation engineering field, such areas as operations planning, logistics, network analysis, financing, and policy analysis are also important to civil engineers, particularly to those working in highway and urban transportation. The National Council of Examiners for Engineering and Surveying (NCEES) list online the safety protocols, geometric design requirements, and signal timing.

Transportation engineering, as practised by civil engineers, primarily involves planning, design, construction, maintenance, and operation of transportation facilities. The facilities support air, highway, railroad, pipeline, water, and even space transportation. The design aspects of transport engineering include the sizing of transportation facilities (how many lanes or how much capacity the facility has), determining the materials and thickness used in pavement designing the geometry (vertical and horizontal alignment) of the roadway (or track). Before any planning occurs the Engineer must take what is known as an inventory of the area or if it is appropriate, the previous system in place. This inventory or database must include information on (1) population, (2)l and use, (3) economic activity, (4) transportation facilities and services, (5) travel patterns and volumes, (6) laws and ordinances, (7) regional financial resources, (8) community values and expectations. These inventories help the engineer create business modells to complete accurate forecasts of the future conditions of the system Review.

Operations and management involve traffic engineering, so that vehicles move smoothly on the road or track. Older techniques include signs, signals, markings, and tolling. Newer technologies involve intelligent transportation systems, including advanced traveller information systems (such as variable message signs), advanced traffic control systems (such as ramp metres), and vehicle infrastructure integration. Human factors are an aspect of transport engineering, particularly concerning driver-vehicle interface and user interface of road signs, signals, and markings.

Engineers in this specialisation:

- Handle the planning, design, construction, and operation of highways, roads, and other vehicular facilities as well as their related bicycle and pedestrian realms.
- Estimate the transportation needs of the public and then secure the funding for the project.
- Analyse locations of high traffic volumes and high collisions for safety and capacity.
- Use civil engineering principles to improve the transportation system.
- Utilises the three design controls which are the drivers, the vehicles, and the roadways themselves.

Railroad Engineering

Railway engineers handle the design, construction, and operation of railroads and mass transit systems that use a fixed guideway (such as light rail or even monorails). Typical tasks would include determining horizontal and vertical alignment design, station location and design, and construction cost estimating. Railroad engineers can also move into the specialised field of train dispatching which focuses on train movement control.

Railway engineers also work to build a cleaner and safer transportation network by reinvesting and revitalising the rail system to meet future demands. In the United States, railway engineers work with elected officials in Washington, D.C. on rail transportation issues to make sure that the rail system meets the country's transportation needs.

Port and Harbour Engineering

Port and harbour engineers handle the design, construction, and operation of ports, harbours, canals, and other maritime facilities. This is not to be confused with marine engineering.

Airport Engineering

Airport engineers design and construct airports. Airport engineers must account for the impacts and demands of aircraft in their design of airport facilities. These engineers must use the analysis of predominant wind direction to determine runway orientation, determine the size of runway border and safety areas, different wing tip to wing tip clearances for all gates and must designate the clear zones in the entire port.

Municipal or Urban Engineering

Municipal or urban engineering applies the tools of science, art and engineering in an urban environment. Municipal engineering is concerned with municipal infrastructure. This involves specifying, designing, constructing, and maintaining streets, sidewalks, water supply networks, sewers, street lighting, municipal solid waste management and disposal, storage depots for various bulk materials used for maintenance and public works (salt, sand, etc.), public parks and bicycle paths. In the case of underground utility networks, it may also include the civil portion (conduits and access chambers) of the local distribution networks of electrical and telecommunications services. It can also include the optimising of garbage collection and bus service networks. Some of these disciplines overlap with other civil engineering specialities, however municipal engineering focuses on the coordination of these infrastructure networks and services, as they are often built simultaneously (for a given street or development project), and managed by the same municipal authority.

History

Modern municipal engineering finds its origins in the 19th century United Kingdom, following the Industrial Revolution and the growth of large industrial cities. The threat to urban populations from epidemics of waterborne diseases such as cholera and typhus lead to the development of a profession devoted to "sanitary science" that later became "municipal engineering".

A key figure of the so-called "public health movement" was Edwin Chadwick, author of the parliamentary report "The Sanitary Condition of the Labouring Population", published in 1842.

Early British legislation included:

- Burgh Police Act 1833 - powers of paving, lighting, cleansing, watching, supplying with water and improving their communities.
- Municipal Corporations Act 1835
- Public Health Act 1866 – formation of drainage boards
- Public Health Act 1875 known at the time as the Great Public Health Act.

This legislation provided local authorities with powers to undertake municipal engineering projects and to appoint borough surveyors (later known as "municipal engineers").

In the U.K, the Association of Municipal Engineers, (subsequently named Institution of Municipal Engineers), was established in 1874 under the encouragement of the Institution of Civil Engineers, to address the issue of the application of sanitary science. By the early 20th century, Municipal Engineering had become a broad discipline embracing many of the responsibilities undertaken by local authorities, including roads, drainage, flood control, coastal engineering, public health, waste management, street cleaning, water supply, sewers, waste water treatment, crematoria, public baths, slum clearance, town planning, public housing, energy supply, parks, leisure facilities, libraries, town halls and other municipal buildings.

In the UK, the development of different strands of knowledge necessary for the management of municipal infrastructure led to the emergence of separate specialised institutions, including:

- For drainage: Chartered Institution of Water and Environmental Management, 1895
- For town planning: Town Planning Institute 1914 ... subsequently becoming the Royal Town Planning Institute
- For street lighting: Association of Public Lighting Engineers, 1934 subsequently becoming the Institution of Lighting Engineers
- For highway engineering: Institution of Highways and Transportation, 1930
- For public housing: Institute of Housing, 1931.

In 1984 the Institution of Municipal Engineers merged with the Institution of Civil Engineers. Since the 1970's, there has been a global trend toward increasing privatisation and outsourcing of municipal engineering services. In the UK in the 1990's a change in management philosophy brought the demise of the traditional organisational structure of boroughs where the three functions of town clerk, borough treasurer and borough engineer were replaced by an administrative structure with a larger number of specialised departments. In the late 1990's and early 21st century there was increasing dissatisfaction over what was perceived to be fractured and dysfunctional public services designed along narrow specialities. A more holistic approach to urban engineering began to emerge as an alternative concept. Critics of the specialised approach included the Commission for Architecture and the Built Environment that complained that the specialised approach to management of the public realm focused too much on the on efficient movement of vehicles rather than the more general interests of local communities.

Professional Practice

In the United Kingdom there is no longer any formal professional qualification in municipal engineering although there are degree courses available in urban engineering. A professional certificate in Urban Engineering is available from through the Institution of Incorporated Engineers via the Public Realm Information and Advice Network.

The British Institution of Civil Engineers (ICE) caters to practitioners employed in the public sector, private consultancy and academia through its Proceedings Journal Municipal Engineer. The journal, first published in 1873, has a global scope and covers the whole life cycle of municipal services addressing technical, political and community issues. In addition an Expert Panel responds on behalf of ICE to Government consultations and is represented on the International Federation of Municipal Engineering.

International Organisation

The International Federation of Municipal Engineering (IFME) is an organisation comprising professional municipal engineers from all round the world. IFME's mission is to connect municipal engineers, public works professionals, public agencies, institutions and businesses around the world in order that they can share a global pool of knowledge and experience. The aim is to foster continued improvement in the quality of public works and wider community services.

The inaugural meeting was held in 1960 at the UNESCO headquarters in Paris. Membership has grown steadily and in 2009 comprised representatives from national associations in: Australia, Canada, Denmark, Estonia, Finland, Italy, Israel, The Netherlands, New Zealand, Norway Southern Africa, Sweden and the UK. Belgium and San Marino are presently Corresponding Members.

Related Engineering Disciplines

Municipal or urban engineering combines elements of environmental engineering, water resources engineering and transport engineering.

Relationship to Urban Design or Urban Planning

Today, municipal engineering may be confused with urban design or urban planning. Whereas the urbanist or urban planner may design the general layout of streets and public places, the municipal engineer is concerned with the detailed design. For example, in the case of the design of a new street, the urbanist may specify the general layout of the street, including landscaping, surface finishings and urban accessories, but the municipal engineer will prepare the detailed plans and specifications for the roads, sidewalks, municipal services and street lighting. However Municipal Engineering as practised a century ago fully embraced the function of urban design and urban planning, even though the terms had yet to be coined.

Traffic Engineering (Transportation)

Traffic engineering is a branch of civil engineering that uses engineering techniques to achieve the safe and efficient movement of people and goods on roadways. It focuses mainly on research and construction of the infrastructure necessary for safe and efficient traffic flow, such as road geometry, sidewalks and crosswalks, segregated cycle facilities, shared lane marking, traffic signs, road surface markings and traffic lights.

Traffic engineering is closely associated with other disciplines:

- Transport engineering
- Pavement engineering
- Bicycle transportation engineering
- Highway engineering
- Urban planning.

Typical traffic engineering projects involve designing traffic control device installations and modifications, including traffic signals, signs, and pavement markings. However, traffic engineers also consider traffic safety by investigating locations with high crash rates and developing countermeasures to reduce crashes. Traffic flow management can be short-term (preparing construction traffic control plans, including detour plans for pedestrian and vehicular traffic) or long-term (estimating the impacts of proposed commercial developments on traffic patterns). Increasingly, traffic problems are being addressed by developing systems for intelligent transportation systems, often in conjunction with other engineering disciplines, such as computer engineering and electrical engineering.

Traffic Systems

Traditionally, road improvements have consisted mainly of building additional infrastructure. However, dynamic elements are now being introduced into road traffic management. Dynamic elements have long been used in rail transport. These include sensors to measure traffic flows and automatic, interconnected, guidance systems to manage traffic (for example, traffic signs which open a lane in different directions depending on the time of day). Also, traffic flow and speed sensors are used to detect problems and alert operators, so that the cause of the congestion can be determined, and measures can be taken to minimize delays. These systems are collectively called intelligent transportation systems.

Lane Flow Equation

The relationship between lane flow (Q, vehicles per hour), maximum speed (V, kilometres per hour) and density (K, vehicles per kilometre) is:

$$Q = KV$$

Observation on limited access facilities suggests that up to a maximum flow, speed does not decline while density increases. However, above a critical threshold, increased density reduces speed. Additionally, beyond a further threshold, increased density reduces flow as well. Therefore, speeds and lane flows at bottlenecks can be kept high during peak periods by managing traffic density using devices that limit the rate at which vehicles can enter the highway. Ramp metres, signals on entrance ramps that control the rate at which vehicles are allowed to enter the mainline facility, provide this function (at the expense of increased delay for those waiting at the ramps).

Highway Safety

Highway safety engineering is a branch of traffic engineering that deals with reducing the frequency and severity of crashes. It uses physics and vehicle dynamics, as well as road user psychology and human factors engineering, to reduce the influence of factors that contribute to crashes.

A typical traffic safety investigation follows these steps

1. Identify and priorities investigation locations. Locations are selected by looking for sites with higher than average crash rates, and to address citizen complaints.
2. Gather data. This includes obtaining police reports of crashes, observing road user behaviour, and collecting information on traffic signs, road surface markings, traffic lights and road geometry.
3. Analyse data. Look for collisions patterns or road conditions that may be contributing to the problem.
4. Identify possible countermeasures to reduce the severity or frequency of crashes.
    - Evaluate cost/benefit ratios of the alternatives
    - Consider whether a proposed improvement will solve the problem, or cause "crash migration." For example, preventing left turns at one intersection may eliminate left turn crashes at that location, only to increase them a block away.
    - Are any disadvantages of proposed improvements likely to be than the problem you are trying to solve?

5. Implement improvements.
6. Evaluate results. Usually, this occurs some time after the implementation. Have the severity and frequency of crashes been reduced to an acceptable level? If not, return to step 2.

Bicycle Transportation Engineering

Bicycle transportation engineering is the study of transportation engineering as it affects bicycles and cycling. It includes the design of dedicated transport facilities for cyclists, but also the study of how mixed-mode environments (i.e. where cyclists are not segregated from other traffic) can be made to work safely.

Roads

Various methods of altering or reallocating the roadway right-of-way to facilitate bicycling and create bikeways have been added to many of the manuals used by transport planners and engineers.

Segregated Cycle Facilities

Sidepath or Shared use Path: A shared use path is a bikeway that is physically separated from motorised vehicle traffic by an open space (generally landscaping) or some form of barrier.

Cycle Track

A cycle track, also called a *separated cycle lane,* is a bicycle exclusive facility that provide physical separation from motorised vehicle traffic within the road right-of-way, generally by means of kerbs or other barriers. Cycle tracks can either incorporate bicycle-only signal phases at intersections (for 100% separation) or utilise "mixing zones" to merge bicycle and motor vehicle traffic. A cycle track combines the user experience of a separated path with the on-road infrastructure of a bike lane.

Bicycle Lane

A designated bicycle lane, according to the 1998 United States Manual on Uniform Traffic Control Devices, is:

- a portion of a roadway which has been designated for use by bicyclists
- marked by a white (usually solid) stripe painted on the pavement
- significantly narrower than traffic lanes
- found at the side of the traffic lanes.

Cycle lanes require less space compared with the option of providing side paths or separate tracks, and generally do not have the legal / priority issues that facilities separated from the general traffic may have (depending on the locally applicable road laws). However, if traffic is heavy or fast, the lack of physical separation from motor vehicles may make cycle lanes unattractive to less confident cyclists. The lack of separation also means they are more apt to be blocked by (illegally) parked vehicles.

Shoulder

A hard shoulder, or simply shoulder, is a reserved area outside of a roadway, but within the road right-of-way. In addition to (a) serving as a buffer area in the event of a temporary loss of vehicle control, (b) providing emergency access for ambulances and police cars and (c) providing a space for inoperable vehicles, it may also be used by pedestrians and bicyclists.

Non-segregated Facilities

Wide Outside Lane (WOL): Wide outside through lanes help motorists pass slower cyclists without having to decrease speed or change lanes. This method is held by some to be particularly important on roads with a high proportion of wide vehicles, such as buses or heavy trucks. These lanes also provide more room for cyclists to filter past queues of other vehicles in congested conditions. The use of such lanes is specifically endorsed by *Cycling: the way ahead for towns and cities*, the European Commission policy document on cycle promotion.

Cycle friendly infrastructure argues for a marked lane width of 4.25 m (14 ft). It is argued that, on undivided roads, this width provides cyclists with adequate clearance from passing wide vehicles while being sufficiently narrow to deter motorists from attempting to "double up" and form two lanes. This "doubling up" effect may be related to junctions. At non-junction locations, greater width might be preferable if this effect can be avoided.

Shared Bus Lane

A shared bus lane is a bus lane that allows cyclists to use it. Depending on the width of the lane, the speeds and number of buses, and other local factors, the safety and popularity of this arrangement vary. Guidance produced for Cycling England endorses bus lanes as providing cyclists with a *direct and barrier free route into town centres* and as avoiding the difficulties associated with other provisions such as shared-use footways. According to a French survey 42% of cyclists described themselves as "enthusiasts" for shared bus bike lanes versus 33% who were of mixed opinion and 27% who were opposed. Many cycling activists view these as being more attractive than cycle paths, while others object to being in close proximity to bus exhausts. The sharing of bus lanes has been described as "generally very popular" with cyclists in London.

As of 2003, mixed bus/cycle lanes accounted for 118 km of the 260 km of cycling facilities in Paris. The French city of Bordeaux has 40 km of shared bus cycle lanes. It is reported that that in the city of Bristol, a showcase bus priority corridor, where road space was reallocated along a 14 km stretch also resulted in more space for cyclists and had the effect of increasing cycling. The reverse effect has also been suggested, a review carried out in London reports that cycling levels fell across Kew bridge following the removal of a bus lane - this was despite a general increase in cycling level in the city generally. In addition, it is arguably easier, politically speaking, to argue for funding of joint facilities rather than the additional expense of both segregated cycling facilities and bus-only lanes.

In some instances bus lane proposals have run into vehement opposition from cycling representatives - a typical theme is the perceived generation of conflict due to the narrowing of other lanes already shared by cars/cyclists so as to create space for the bus lanes The TRL reports that cyclists and bus drivers tend to have low opinions of each other There have been reports in Dublin of conflict as cyclists choose to cycle in the bus lanes and a bus driver apparently expected them to use adjacent cycle tracks instead. In some other cities the arrangements seem to work successfully, with bus companies and cyclists' groups taking active steps to ensure that understanding is improved between the two groups of road users.

Streets

Shared Space: Shared space schemes, which are characterised by the removal of road markings, signs and signals, give all street users equal priority and equal responsibility for each others safety. Experiences where these schemes are in use, show that street users, particularly motorists, undirected by signs, curbs or road markings, reduce their speed and establish eye contact with other users. Results from the thousands of such implementations worldwide all show casualty reductions and most also show reduced journey times. Following the partial conversion of London's Kensington High Street to *shared space,* accidents were reduced there by 44% (the London average was 17%).

Bicycle Boulevards

A bicycle boulevard is a low speed street which has been designed to discourage cut-through motor vehicle traffic and to give priority to cyclists as through-going traffic.

Bike Paths

Bike paths are bicycle routes that follow an independent right-of-way. There are two distinct types of bike paths: those used exclusively by bicycles and those shared with pedestrians. In the United States almost all bike paths are shared with pedestrians.

Bike paths are often built on old railroad right-of-ways (called rail trails) and as part of greenways and fore shoreways.

Traffic Lights

How traffic signals are designed and implemented directly impacts cyclists. For instance, poorly adjusted vehicle detector systems, used to trigger signal changes, may not correctly detect cyclists. Traffic managers in Copenhagen link cyclist-specific traffic signals on a major arterial bike lane to provide green waves for rush hour cycle-traffic. Cycling-specific measures that can be applied at traffic signals include the use of advanced stop lines and/or bypasses.

The frequency with which lights change is important to cyclists who may conserve energy by anticipating green lights ahead i.e. the shorter the interval the better for cyclists. Intersection clearance times and green wave timing may also be relevant.

Road Surface

Bicycle tires being narrow, road surface is more important than for other transport, for both comfort and safety. The type and placement of storm drains, manholes, surface markings, and the general road surface quality should all be taken into account by a bicycle transportation engineer. Drain grates, for example, must not catch wheels.

Parking

Bicycle parking is another important part of Bicycle Transportation Engineering. In most of the United States, bicycle parking facilities are scarce, or are so inadequate that nearby trees or parking metres are used. The hitching post type of bicycle stand is an improvement over the old type that had a slot for the front wheel but only allow for two bicycles per post. The Netherlands, where bicycles are much in use, has two-tiered bicycle racks giving high density (the handlebars overlap) and security (the bicycle is held well and is easy to lock).

Secure bicycle parking is argued to be a key factor influencing the decision to cycle. To be considered secure, the parking must be of a suitable design: allowing the bicycle to be locked via the frame. A readily observable location can also permit so-called passive security from passers-by. Weather protection is also desirable. As a rule, where cycling is encouraged as an alternative to motoring, efforts are made to make bicycle parking more convenient and attractive to use than nearby car parking arrangements. This usually means providing a wide distribution of visible, clearly designated parking spots, close to the entrances of destinations being served.

Storage rooms or bicycle lockers may also be provided. In some cases large concentrations of bike parking may be more appropriate, sometimes being supervised and sometimes charging a fee. Examples include large bike parking facilities at public transport interchanges such as railway, subway, tram, bus stations, or at bike stations where they may be useful in Mixed-mode commuting.

Conversely, where cycling is seen as an unwelcome or inappropriate activity, or due to lack of knowledge about best practices, bicycle parking may simply not be provided or else placed at awkward, distant, and out-of-sight locations. Cyclists may be expressly forbidden from parking their bicycles at the most convenient locations. In April 2007, the authorities at the University of California's Santa Barbara campus started confiscating bicycles not parked at the allegedly inconvenient official bike racks. Often, property owners display signage on fencing to discourage bicyclists from locking their bicycles.

Pavement Engineering

Pavement engineering is a branch of civil engineering that uses engineering techniques to design and maintain flexible (asphalt) and rigid (concrete) pavements.

This includes streets and highways and involves knowledge of soils, hydraulics, and material properties. Pavement engineering involves new construction as well as rehabilitation and maintenance of existing pavements. Maintenance often involves using engineering judgment to make maintenance repairs with the highest long-term benefit and lowest cost. The Pavement Condition Index (PCI) is an example of an engineering approach applied to existing pavements. Another example is the use of a falling weight deflectometre (FWD) to non-destructively test existing pavements. Calculation of pavement layer strengths can be performed from the resulting deflection data.

Four on the Floor! (Strength vs. Stability)

There are two issues that will be emphasized throughout this course that are crucial to the understanding of Architectonics:

- Strength : the capacity of the individual elements, which together make up a structural system, to withstand the load that is applied to it.
- Stability: the capability of a structural system to transmit various loadings safely to the ground.

These two critical issues are experienced daily from the moment that an individual is born. A newborn baby cannot even hold its own head upright. The large mass of the head requires a support system that has sufficient strength to enable the head to maintain its stability. This steadily increases as the bones, muscles and tendons of the skeletal and muscular systems increase in strength. Eventually the extra support provided by the arm or hand is no longer needed. The first challenge posed by gravity is overcome.

Crawling on four points of support proves to be a very stabile situation for quite a long time. The "leap" to the unstable two point stance is the next development in our understanding of the influence of gravity. Again, the structural system must develop to the point that the individual elements of the system have acquired sufficient strength. The first steps are made: an action of supreme coordination of hundreds of elements that becomes second nature to homo sapiens. The list can be extrapolated to touch on many aspects of the human experience; riding tricycles and bicycles, jumping on trampolines, exercising on parallel bars, sliding on ice skates, sailing in a heavy wind, rocking a small boat, the list is endless. These are part of the human experience and each and every one rely on an inherent understanding of strength and stability.

How many times has a parent scolded a child to "put four on the floor!!!"? What the parent really means to say is, "if you do not put all of the legs of your chair on the ground, you are going to tip over!" Both strength and stability issues are addressed in this simple exclamation. Under normal conditions, the elements which make up the chair (its legs, bracing and seat) can easily resist the implied vertical loads. The strength of the individual elements of the chair have been designed for this type of static load. The seat (as a horizontal load-bearing element) must transfer its load through a connection to the legs (vertical load-bearing elements). Granted, some chairs will withstand a greater load than others, but they all resist the pull of gravity on the person sitting in them. If the legs cannot support the applied load they will fracture or break. These are examples of strength failure.

The stability of the system of elements depends upon the orientation of the chair in space. When it stands upright, on all four legs, it is a stable system. If it is on it's side, the chair might not be able to resist the loads for which it was designed. As it is tilted onto the back two legs, the structural system loses its equilibrium. At a certain point the chair as a system becomes unstable, fails and gravity pulls the supported load to the ground. This is a stability failure. In this type of failure, the individual elements retain their strength even as the system fails. The chair (system) could also have failed if the two supporting legs had experienced a strength failure (broken).

In each of these situations the chair, as a structural system, has reached the limit of its strength. As the saying goes, a chain (structural system) is only as strong as the weakest link (element)!

Any structural system can be studied in light of these two issues. For example, the column of the Greek temple shown above is an element that can experience a strength (crushing) failure, or a system (buckling) failure. It is/was part of a larger structural system.

The Role of Structural Failure

The concepts of element strength and system stability are learned at an early age. They often appear in children's stories that have been repeated for decades. One favourite in North America is the story of the "Three Little Pigs." Each pig desired to build a house to protect itself from the Big, Bad Wolf. One pig built a house of straw while another built a house of sticks and twigs. The third pig erected a house of brick. The story follows that the big, bad wolf came to the homes of the three pigs. He drew a huge breath and huffed and puffed until he blew the first house down. He chased the pig into the second house where he repeated his imitation of a typhoon. This house was also blown over and the pigs fled into the brick house. In front of the brick house the wolf hyperventilated. His grandest attempts to blow over this ediface failed and so he did not have pork for dinner. The story teaches children that a brick building is the strongest and most stable becuase it is the only one which could resist the huffing and puffing. The straw and stick buildings were blown away in the wind and only within the brick structure do the three pigs find their security. This lesson is correct if the load that is of greatest concern is a tornado, typhoon, hurricane or similar lateral load. However, the brick house would have been the worst structure to live in if it was built in an earthquake zone. In fact, an unreinforced masonry building is much more life threatening in an earthquake when compared to a straw hut. Then again, another response would be appropriate if the house was in a region where winds were not a real problem, but extreme temperature variations were critical. The same would be true if mobility was the most significant issue. Each of these are specific loading conditions that could lead to structural failure. It is critical that the appropriate loads and loading combinations are considered for each and every structure depending on use and location. Every one is different!

"Humpy Dumpty" is another childrens' tale with a structural lesson. He was a good egg who sat on a wall for quite a while. One day he lost his balance and fell to the ground where all of the king's men could not put Humpty back together again. The egg shell is a wonderfully efficient, naturally occurring thin structure that is extremly strong under evenly distributed loading conditions. One type of load which a thin shell cannot resist is a sharp point load. Thus, when Humpty fell to the ground he loaded his wonderful structure in a way that caused a strength and system failure. Building Codes are often considered to be the bane of architectural designers. However, they have been a part of the profession for thousands of years. The illustration above is a part of the Code of Hammurabi that was written about 2200BC. The Code can be translated as:

If a builder builds a house for a man and does not make its construction firm and the house collapses and causes the death of the owner of the house - that builder shall be put to death. If it causes the death of a son of the owner - a son of that builder shall be put to death.

If it causes the death of a slave of the owner - the builder shall give the owner a slave of equal value. If it destroys property - the builder shall restore whatever it destroyed and because the builder did not make the house firm, shall rebuild the house which collapsed at his own expense.

If a builder builds a house and does not make its construction meet the requirements and a wall falls in - that builder shall strengthen the wall at his own expense.

The negative incentive of the Code of Hammurabi does not bode well for innovation. However, it clearly demonstrates the point that building codes have existed for a long time, and are not likely to go away. Structural designers are often considered to be a very conservative group. One of the reasons for this could be found in such a building code. In addition, when a building experiences structural failure the results are usually quite different than if the air conditioning system fails. Although some aspects of the building codes seem ridiculous, they are in place to guarantee the public's safety and should be respected.

Each and every one of us is a structural designer in their own way. The act of walking down a set of stairs or pushing a wheelchair forward or swimming across a pool requires an understanding of gravity, the flow of forces and how to work with them.

Petroski notes: "The Concept of Failure is central to understanding engineering, it has as its first and foremost objective the obviation of failure. Design, even structural design, is a human endeavour and thus it is subject to error. Due to this, some designs are destined to fail. This can lead to a loss of life which in itself is tragic, but a deeper unforgivable tragedy exists when the lessons of the failure are understood and allowed to occur again."

What is a Force?

The one constant around the world is the action of gravity upon each and every structure that is erected. The primary function of all structural design is to make a building stand-up. Understanding architectonics will enable a designer to include these issues as part of a design language that will create a significantly clearer architectural expression. Primary to this study is the concept of a force. A force is actually a very abstract conception. It can be defined, but it cannot become physically apparent until it meets resistance. Imagine a six foot tall block of ice sliding along a frictionless surface laid inside of a hockey rink. If a person props herself against the wall and tries to stop the ice she will then perceive the force imparted by the block of ice. She transfers the force of the block of ice into a force that moves into the ground. Thus, if the ground could actually experience a push, it would as well as the slide is arrested.

A "force" is an action that changes, or tends to change, the state of motion of the body upon which it acts. It is a vector quantity that can can be represented either mathematically or graphically. A complete description of a force MUST include its:

1. Magnitude
2. Direction qnd Sense
3. Point of Application.

The Magnitude is most often expressed in the units of pounds (lbs), newtons (N), kilo-newtons (KN) or kilo-pounds (KIPS; 1 Kip = 1000 lbs). The magnitude is represented graphically by the scaled length of the arrow which represents the force. Graphic statics depends upon the accurate representation of the magnitude of each force acting upon a body.

The Direction of a force is discerned by oberserving the line of action of the vector of the force. The Sense refers to the direction of the movement of the vector along that line of action.

The sense is always represented on the vector by an arrowhead. These two attributes can be either a written verbal description or more conveniently expressed in terms of 360 degrees. In the later case, one begins with zero and increases clockwise with the direction of the arrowhead until 360 is reached.

The sense of the force can also be expressed as a positive or negative sign. This is especially useful when combining forces algebraically.

It is very important not to confuse the direction and sense. The direction always relates to the line of action of the vector, and the sense is the way in which the vector would move along that line. In the example, the sense of the 100 lb force would be "down and to the left" or "210 degrees."

The Point of Application is often overlooked in the description of a force. However, it is fully as important as the magnitude of the force! It is the exact location of the application of a force on a body. It can either be a relative measurement or a set of coordinates.

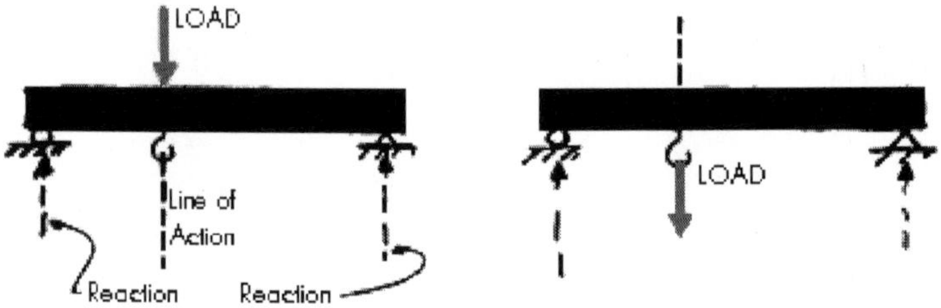

A 100 kip force is applied to the stone column in the diagram. The column will experience this force at every point along the line of action of the force. As a matter of fact, the force will also be transferred to the ground which is supporting the column. Thus, the earth below the column along the line of action of the force will also experience the 100 kip load. This illustrates the Principle of

Transmissibility. The point of application of an external force acting on a body (structure) may be transmitted anywhere along the force's line of action without affecting the other external forces (reactions and loads) acting on that body. This means that there is NO NET CHANGE in the static effect upon any body if the body is in equilibrium. This can be illustrated with the following diagram.

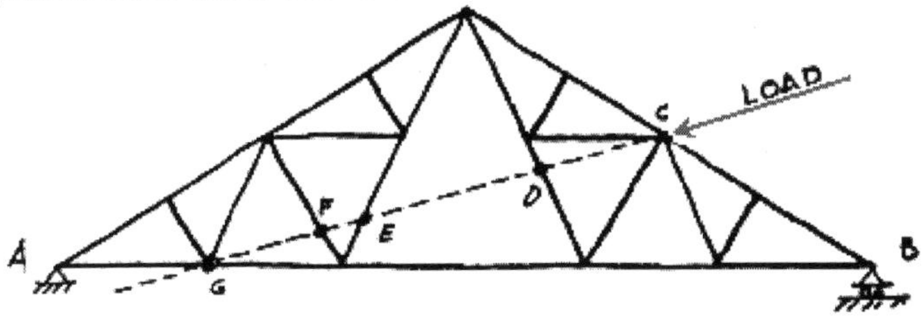

Assume that a beam is supported at its ends. A load is applied to the top of the beam that is acting downward. This load could be a person standing on the beam. The load creates reactions that push up at the two points of support. The line of action of the load (person) on the beam also passes through a hook that is attached to the underside of the beam. Now, if the person standing on top of the beam would climb down and hold on to the hook exactly below the point where they were previously located so that the lines of action were exactly the same, the reactions at the ends of the beams would not change. This is because the load of the person is still acting along the same line of action. As long as a load is applied at any point along the line of action the external reactions will not change.

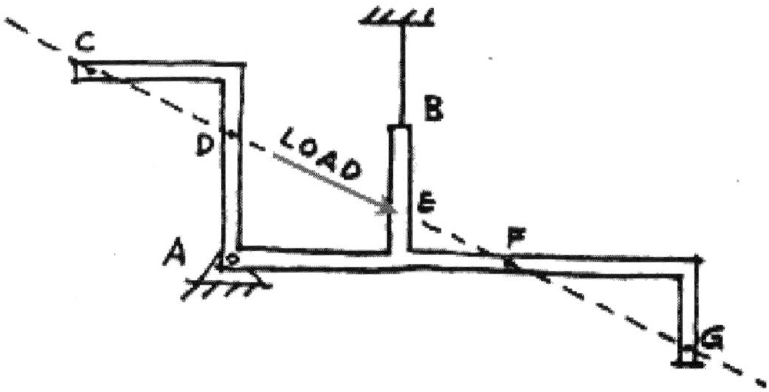

The truss above is loaded with a force that is applied at point C. This load creates reactions at the two supports A and B. The load on the truss could move anywhere along the line of action and the external reactions at A and B would remain the same. That means that if the load was applied at points D, E, F or G the reactions at

A and B would not change. Note that the only point of discussion at this moment is the fact that the external reactions will not change. It is clear that the internal forces will vary greatly within the truss as the force is moved along the line of action. This issue will be discussed in a later lecture.

This strangely shaped beam is another example of a structure that is loaded at a specific point, namely E. In order for the structure to remain at rest there must be reactions of some kind at points A and B. The reaction at B is a tension reaction since it is a cable. Again, if the load was applied at points C, D, E, F or G on this rigid body (structure) the reactions at A and at B would remain exactly the same.

Only the external forces (reactions) remain unchanged. Some of the internal resisting forces within the elements of the structure change as the load is applied at different points along its line of action. This illustrates one important difference between INTERNAL and EXTERNAL forces.

The Principle of Transmissibility applies to any body (blobs, balloons, simple beams, crooked beams, trusses, shells, etc.). It is independent of the body's size or shape.

Moments

The Moment of a force is a measure of its tendency to cause a body to rotate about a specific point or axis. This is different from the tendency for a body to move, or translate, in the direction of the force. In order for a moment to develop, the force must act upon the body in such a manner that the body would begin to twist. This occurs every time a force is applied so that it does not pass through the centroid of the body. A moment is due to a force not having an equal and opposite force directly along it's line of action.

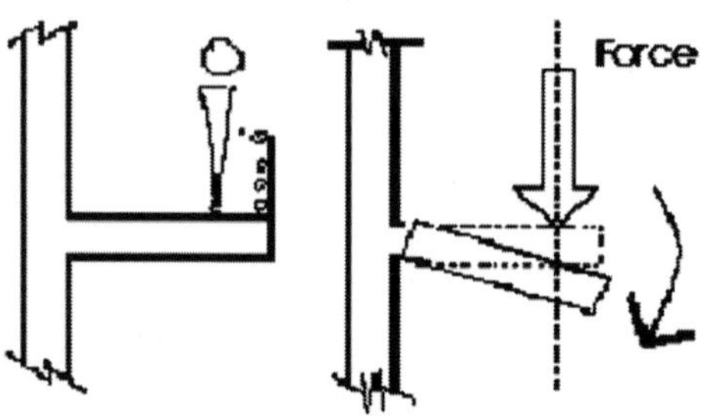

Imagine two people pushing on a door at the doorknob from opposite sides. If both of them are pushing with an equal force then there is a state of equilibrium. If one of them would suddenly jump back from the door, the push of the other person would no longer have any opposition and the door would swing away. The person who was still pushing on the door created a moment.

The magnitude of the moment of a force acting about a point or axis is directly proportional to the distance of the force from the point or axis. It is defined as the product of the force (F) and the moment arm (d). The moment arm or lever arm is the perpendicular distance between the line of action of the force and the centre of moments.

The Centre of Moments may be the actual point about which the force causes rotation. It may also be a reference point or axis about which the force may be considered as causing rotation. It does not matter as long as a specific point is always taken as the reference point. The latter case is much more common situation in structural design problems. A moment is expressed in units of foot-pounds, kip-feet, newton-metres, or kilonewton-metres. A moment also has a sense; it is either clockwise or counter-clockwise. The most common way to express a moment is :

$$Moment = Force \times Distance$$

$$M = F \times d$$

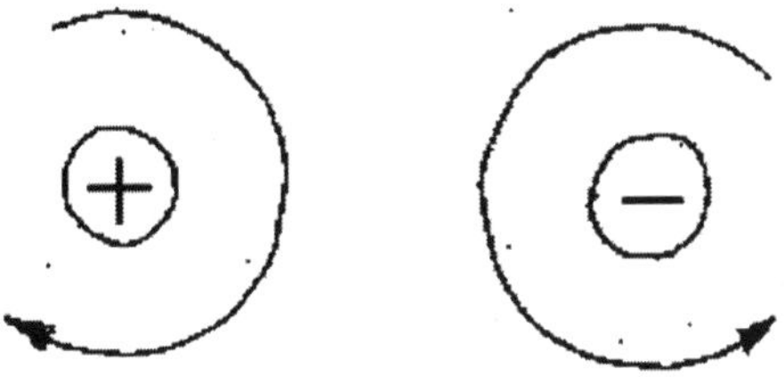

The example shown is a wrench on a nut (at C) that has a force applied to it. The force is applied at a distance of 12 inches from the nut. The centre of moments could be point C or points A or B. The moment arm for calculating the moment around point C is 12 inches. The magnitude of the moment of the 100 pound force about point C is 12 inches multiplied by the force of 100 pounds to give a moment of 1200 inch-pounds (or 100 foot-pounds). Similarly, the moment about point A can be found to be 800 inch-pounds.

The direction of the rotation is important to understand in order to describe its effect on the body (structure). A moment will cause either a clockwise or counter-clockwise rotation about the centre of moments. It is essential that the direction of rotation about the centre of moments be understood. Since an international convention does not exist, in this course, a clockwise rotation about the centre of moments will be considered as positive; a counter-clockwise rotation about the centre of moments will be considered as negative. A moment causes a rotation about a point or axis. Thus, the moment of a force taken about any point that lies on its own line of action is zero. Such a force cannot cause a rotation because the moment arm is nonexistant.

In this second example, a 200 pound force is applied to the wrench. The moment of the 200 pound force applied at C is zero because 200 pounds x 0 inches = 0 inch-pounds (Fxd = M). In other words, there is no tendency for the 200 pound force to cause the wrench to rotate the nut. One could increase the magnitude of the force until the bolt finally broke off. A moment can also be considered to be the result of forces detouring from a direct line drawn between the point of loading of a system and its supports. In this case, the blue force is an eccentric force. In order for it to reach the base of the column, it must make a detour through the beam. The greater the detour, the greater the moment. The most efficient structural systems have the least amount of detours possible. This will be discussed in more detail in Lecture 37 and later courses.

There are cases in which it is easier to calculate the moments of the componenets of a force around a certain point than it is to calculate the moment of the force itself. It could be that the determination of the perpendicular distance of the force is more difficult than determining the perpendicular distance of components of the force. The moment of several forces about a point is simply the algebraic sum of their component moments about the same point. When adding the moments of componenets, one must take great care to be consistent with the sense of each moment. It is often prudent to note the sense next to the moment when undertaking such problems.

Any difficulty with calculating a moment can usually be traced to one of the following:

1. The centre of moments has not been correctly established or clearly understood.
2. The assumed moment arm is not the PERPENDICULAR distance between the line of action of the force and the centre of moments.
3. The direction, or sense, of the rotation has been ignored or misunderstood.

Equilibrium

Architectural structures are normally stationary. Most clients, building officials and designers prefer that their structures remain static rather than move dynamically. There are specific loading conditions which are dynamic loads, but in each and every case a return to a stable and static state is desirable. Such a condition is known as equilibrium.

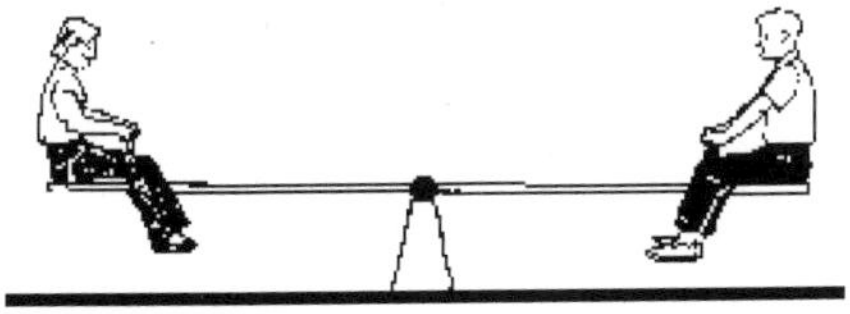

Various states of static equilibrium are experienced throughout one's life. Think of the "teeter-totter" at a playground or of a game of "tug-of-war." In the first case, two or more individuals sit upon a board which has been fixed to a fulcrum which allows rotation. If each of the individuals on the teeter totter weight exactly the same amount and sit at exactly the same distance from the fulcrum the teeter-totter will not move. A state of equilibrium has been achieved. The two will remain at rest until an action takes them out of equilibrium.

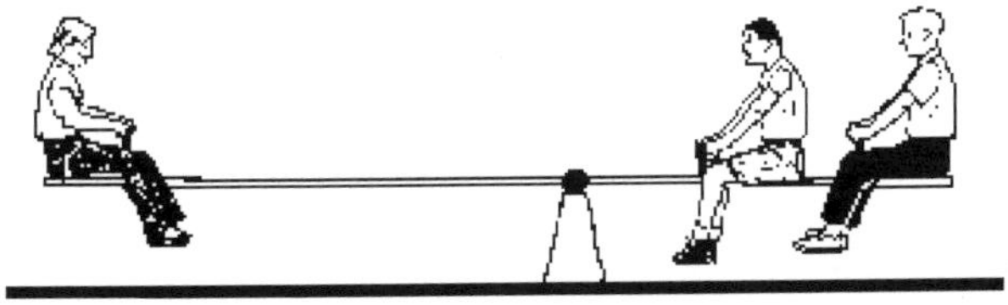

Such an action could be the addition of another person to the system or it could be that one of the original two would change their position slightly. In either case, the teeter-totter would most likely swing to one side and rest upon the ground. A new state of equilibrium would have been found. In this case, another boy climbed on behind the one already sitting on the right. In order to put the system back into equilibrium the girl on the left had to move backwards along the board until she was far away. The moment two boys created around the fulcrum doubled when the second boy climbed on. The girl knew that the only way that she could increase the magnitude of the moment she created would be to increase the moment arm. Thus she moved back until she was twice as far from the fulcrum. Now the system would be back in equilibrium.

Another example of a state of equilibrium is the game of "tug-of-war." A rope is pulled taught between two teams; each pulls with a force that equals the force of the opposite team. Assume in the figure that each team is pulling with a force of 220 pounds. As long as each team maintained a pull of 220 pounds the system is in equilibrium. If during this time a device would be inserted between the two teams to measure the magnitude of the tension force that the rope has anywhere along its length, it would read 220 pounds at each and every point. This would be true at ANY point along the rope. A structure is in equilibrium when all forces or moments acting upon it are balanced. This means that each and every force acting upon a body, or part of the body, is resisted by either another equal and opposite force or set of forces whose net result is zero. Issac Newton addressed this issue when he noted that a body is at rest will remain at rest until acted upon by an external force. Every structure that can be seen to remain standing on a daily basis is in equilibrium; it is at rest and each of its members, combination of its members or any part of a member that is supporting a load are also at rest. There is a net result of zero in all directions for all of the applied loads and reactions.

The beam of the top figure illustrates the system at rest under it's own weight. The lower figure shown another state in which a pair of scissors sits at one end and a thermos of coffee on the opposite side near the fulcrum. All of the forces and moments are balanced so that the system is in a stable equilibrium.

There are two types of equilibrium; External and Internal. External equilibrium encompasses the loads upon, and reactions of, a structural system as a whole. Internal equilibrium describes the various forces the are acting within every member of the system. There are conditions of equilibrium that must be satisfied for each case. These are:

Sum of All Vertical Forces (Fy) = 0

Sum of All Horizontal Forces (Fx) = 0

Sum of All Moments (Mz) = 0

(Sum of All Forces (Fz) = 0)

(Sum of All Moments (My) = 0)

(Sum of All Moments (Mx) = 0)

These six equations are all that can be used to determine every one of the forces that are acting with a structure. They are few, but very powerful. The first three are the most common equations and will be utilised in all of the problems associated with third course. The other three are only necessary when considering three-dimensional force systems.

Bibliography

Armstrong, J. A.: *Forces Shaping the U.S. Academic Engineering Research Enterprise*, National Academy Press, Washington DC, 1995.

Armytage, W. H. G.: *A Social History of Engineering*. Cambridge: MIT Press, 1966.

Beaty H. Wayne: *Standard Handbook for Electrical Engineers*. New York, NY: McGraw-Hill, 2000.

Benjamin Henry: *The Engineering Drawings of Benjamin Henry Latrobe*, Yale University Press, NY, 1980.

Bennett, S.: *A History of Control Engineering: 1800-1930*, IEE Press, London, 1979.

Bhavikatti, S. S. : *A Textbook on Elements of Civil Engineering and Engineering Mechanics*, New Age International , Delhi, 2011.

Booker, P. J.: *A History of Engineering Drawing*, Chatto & Windus, London, 1963.

Calhoun, Daniel H. : *The American Civil Engineer: Origins and Conflict*. Cambridge: MIT Press, 1960.

Chen, W.F.: *Civil Engineering Handbook*. Boca Raton, FL: CRC Press, 1995.

David Fisher: *Rules of Thumb for Engineers and Scientists*, Gulf Pub, Houston, 1991.

Davison, C. S. C. B.: *Engineering Heritage*, Institute of Mechanical Engineers, London, 1963.

Domminghaus, H.: *Plastics for Engineers: Materials, Properties, Applications*, Hanser Publishers, Munich, 1993.

Dym, J.B.: *Product Design with Plastics: A Practical Manual*, Industrial Press, New York, 1983.

Edward H. Smith: *Mechanical Engineer's Reference Book*, Butterworth-Heinemann, Oxford, 1994.

Emerson, Howard P.: *Origins of Industrial Engineering: The Early Years of a Profession*. Institution of Industrial Engineers, 1988.

Fernandez, Martin : *Dictionary of Civil Engineering*, A P H, Delhi, 2011.

Florman, S.: *The Civilized Engineer*, St. Martin's Press, New York, 1979.

Groover, Mikell P.: *Fundamentals of Modern Manufacturing*, John Wiley & Sons, INC, New York, 2002.

Gupta, J.P. : *A T B of Engineering Mech. and Basic Civil Engineering*, S. Chand Publishing, Delhi, 2011.

Hasan, Syed Danish : *Civil Engineering Materials and Their Testing*, Narosa, Delhi, 2011.

Howard B., Cary,: *Modern Welding Technology*, Pearson Education, Chicago, New Jersey, 2005.

Joshi, Jagat Pati : *Harappan Architecture And Civil Engineering*, Rupa Pub, Delhi, 2008.

Kapoor, R : *Encyclopedia of Civil Engineering*, SBS Pub, Delhi, 2007.

Mendel, Otto: *Practical Piping Handbook*. Tulsa, OK: PennWell Books, 1981.

Nicholas P.: *Handbook of Chemical Engineering Calculations*, McGraw-Hill, New York, 2004.

Punmia, B.C.; Ashok Kr. Jain and Arun Kr. Jain: *Basic Civil Engineering*, Laxmi Publications, Delhi, 2003.

Robert C. Rosaler: *Standard Handbook of Plant Engineering*, McGraw-Hill, New York, 1995.

Singh, Gurcharan : *Standard Handbook of Civil Engineering*, Standard Pub, Delhi, 2003.

Thomas Sixsmith & R. Hanselka: *Handbook of Thermoplastic Piping System Design*. Marcel Dekker Ltd, London, 1997.

Index

❑❑❑